CATALYTIC ANTIBODIES

The Ciba Foundation is an international scientific and educational charity. It was established in 1947 by the Swiss chemical and pharmaceutical company of CIBA Limited—now CIBA-GEIGY Limited. The Foundation operates independently in London under English trust law.

The Ciba Foundation exists to promote international cooperation in biological, medical and chemical research. It organizes about eight international multidisciplinary symposia each year on topics that seem ready for discussion by a small group of research workers. The papers and discussions are published in the Ciba Foundation symposium series. The Foundation also holds many shorter meetings (not published), organized by the Foundation itself or by outside scientific organizations. The staff always welcome suggestions for future meetings.

The Foundation's house at 41 Portland Place, London W1N 4BN, provides facilities for meetings of all kinds. Its Media Resource Service supplies information to journalists on all scientific and technological topics. The library, open five days a week to any graduate in science or medicine, also provides information on scientific meetings throughout the world and answers general enquiries on biomedical and chemical subjects. Scientists from any part of the world may stay in the house during working visits to London.

Ciba Foundation Symposium 159

CATALYTIC ANTIBODIES

A Wiley-Interscience Publication

1991

JOHN WILEY & SONS

Chichester · New York · Brisbane · Toronto · Singapore

Published in 1991 by John Wiley & Sons Ltd.
Baffins Lane, Chichester
West Sussex PO19 1UD, England

Other Wiley Editorial Offices

John Wiley & Sons, Inc., 605 Third Avenue,
New York, NY 10158-0012, USA

Jacaranda Wiley Ltd, G.P.O. Box 859, Brisbane,
Queensland 4001, Australia

John Wiley & Sons (Canada) Ltd, 22 Worcester Road,
Rexdale, Ontario M9W 1L1, Canada

John Wiley & Sons (SEA) Pte Ltd, 37 Jalan Pemimpin 05-04,
Block B, Union Industrial Building, Singapore 2057

Suggested series entry for library catalogues:
Ciba Foundation Symposia

Ciba Foundation Symposium 159
ix + 259 pages, 66 figures, 10 tables

British Library Cataloguing in Publication Data
A catalogue record for this book is
available from the British Library

ISBN 0 471 92962 X

Phototypeset by Dobbie Typesetting Limited, Tavistock, Devon.
Printed and bound in Great Britain by Biddles Ltd., Guildford.

Contents

Symposium on Catalytic antibodies, held at the Ciba Foundation, London 1–3 October 1990

Editors: Derek J. Chadwick (Organizer) and Joan Marsh

Participants

Y. Arata Faculty of Pharmaceutical Sciences, University of Tokyo, Hongo, Bunkyo-ku, Tokyo 113, Japan

S. J. Benkovic Department of Chemistry, College of Science, 152 Davey Laboratory, The Pennsylvania State University, University Park, PA 16802, USA

G. M. Blackburn Department of Chemistry, University of Sheffield, Sheffield S3 7HF, UK

K. Brocklehurst Department of Biochemistry, Queen Mary & Westfield College, Mile End Road, London E1 4NS, UK

D. R. Burton Department of Molecular Biology, University of Sheffield, Sheffield S10 2TN, UK

B. S. Green Department of Pharmaceutical Chemistry, School of Pharmacy, The Hebrew University of Jerusalem, PO Box 12065, IL-91120 Jerusalem, Israel

D. E. Hansen Department of Chemistry, Amherst College, Amherst, MA 01002, USA

T. J. R. Harris Glaxo Group Research, 891 Green Road, Greenford, Middlesex UB6 0HE, UK

D. Hilvert Departments of Molecular Biology & Chemistry, Research Institute of Scripps Clinic, 10666 North Torrey Pines Road, La Jolla, CA 92037, USA

W. Huse Ixsys Inc, 3550 General Atomics Court, Suite L103, San Diego, CA 92121, USA

B. L. Iverson Department of Chemistry, University of Texas at Austin, Austin, TX 78712, USA

K. D. Janda Departments of Molecular Biology & Chemistry (MB-20), Research Institute of Scripps Clinic, 10666 North Torrey Pines Road, La Jolla, CA 92037, USA

W. P. Jencks Graduate Department of Biochemistry, Brandeis University, Waltham, MA 02254, USA

A. S. Kang Departments of Molecular Biology & Chemistry, Research Institute of Scripps Clinic, 10666 North Torrey Pines Road, La Jolla, CA 92037, USA

R. J. Leatherbarrow Department of Chemistry, Imperial College of Science, Technology & Medicine, London SW7 2AY, UK

R. A. Lerner Departments of Molecular Biology & Chemistry, Research Institute of Scripps Clinic, 10666 North Torrey Pines Road, La Jolla, CA 92037, USA

M. T. Martin IGEN Inc, 1530 E Jefferson Street, Rockville, MD 20852, USA

A. Mountain Department of Protein Engineering, Celltech Ltd, 216 Bath Road, Slough, Berks SL1 4EN, UK

M. I. Page Department of Chemical & Physical Sciences, Huddersfield Polytechnic, Queensgate, Huddersfield HD1 3DH, UK

S. Paul Department of Pharmacology, University of Nebraska Medical Center, 600 South 42nd Street, Omaha, NE 68198-6260, USA

A. Plückthun Gen-Zentrum der Universität München, Max-Planck-Institut für Biochemie, Am Klopferspitz, D-8033 Martinsried/München, Germany

P. G. Schultz Department of Chemistry, University of California, Berkeley, CA 94720, USA

A. W. Schwabacher Department of Chemistry, Iowa State University, Ames, IA 50011, USA

K. Shokat Department of Chemistry, University of California, Berkeley, CA 94720, USA

C. J. Suckling Department of Pure & Applied Chemistry, University of Strathclyde, Thomas Graham Building, 295 Cathedral Street, Glasgow G1 1LX, UK

C. M. Tedford Department of Pure & Applied Chemistry, University of Strathclyde, Thomas Graham Building, 295 Cathedral Street, Glasgow G1 1LX, UK

J. Thornton Department of Biochemistry, University College London, Gower Street, London WC1E 6BT, UK

I. A. Wilson Department of Molecular Biology (MB13), Research Institute of Scripps Clinic, 10666 North Torrey Pines Road, La Jolla, CA 92037, USA

Introduction

W. P. Jencks

Graduate Department of Biochemistry, Brandeis University, Waltham, MA 02254-9110, USA

This meeting is being held at a very appropriate time on a subject that has developed rapidly. Many of the people who have participated in this development are present, and there is no better way to bring a subject of this kind into focus than to assemble those who have contributed to it.

The question may arise of what I am doing here, since I have never made a catalytic antibody and have no results to present. The reason is that I wrote a book some time ago (Catalysis in chemistry and enzymology 1969), because I wanted a place where I could look things up. In 1968 we had been thinking about transition states and their stabilization, following, of course, Linus Pauling. A paragraph in my book points out that since enzymes are complementary to transition states, it should be possible to prepare antibodies to transition state analogues that would stabilize transition states and, therefore, catalyse reactions.

If complementarity between the active site and the transition state contributes significantly to enzymic catalysis, it should be possible to make an enzyme by constructing such an active site. One way to do this is to prepare an antibody to a haptenic group which resembles the transition state of a given reaction. The combining site of the antibody should be complementary to the transition state and should accelerate the reaction by binding substrates and forcing them to resemble the transition state. Richard Pincock, who was visiting our department, attempted to prepare antibodies to transition states of the hydrolysis of an *ortho* ester and of a substitution reaction on a sulphonate. The results were entirely negative, for good reason—we had no monoclonal antibodies and we were looking at the wrong type of reaction. One needs a reaction that is susceptible to catalysis and occurs at a reasonable rate in the absence of catalysis. It is also important to have monoclonal antibodies, because they can be obtained in large quantities.

Raso & Stollar (1975) at Tufts University made antibodies to reduced Schiff bases of pyridoxal phosphate and tyrosine, and showed that the antibodies caused a small increase in the rate of transamination of L-tyrosine. This, to my knowledge, was the first experimental indication that such an advantage could be conferred by a specific antibody. Major developments in the field came later, when monoclonal antibodies became available, and were brought about by the people at this symposium.

My own contribution was peripheral, but it illustrates an important principle in science, namely that it is important to be forced to do things you would never do spontaneously. In this instance, Nate Kaplan, who was chairman of the department of Biochemistry at Brandeis University, pointed out to me during 1965 that the next year would be the 25th anniversary of Fritz Lipmann's review on the nature of the 'high energy' phosphate bond (1941), and a book should be published to commemorate this event. He asked me to write a chapter. I told him that I had nothing to say on this subject, but he told me to think of something. I finally wrote about the stabilization of transition states by enzymes and the possibility that compounds resembling transition states would bind to enzymes very well. The term 'entropic distortion' was used at that time to indicate that the free energy of binding could be utilized to overcome entropy loss, and that a suitable molecule could serve as a type of transition state analogue, and hence give us a model to probe the mechanism of catalysis (Jencks 1966).

All of this went back to Linus Pauling who had published a number of important papers in *Chemical and Engineering News* and the *American Scientist* in the 1940s, which I had never seen. In those papers Pauling pointed out that enzymes must stabilize transition states and that non-covalent factors, such as strain and ground state destabilization, are important for catalysis. Pauling was thus the father of the concepts behind the development of catalytic antibodies, as of so many other developments in chemistry.

We might ask what a catalytic antibody is best fitted to do. The catalytic effect that has probably been thought about most is geometric distortion. The transition state has a different geometry from the ground state, and an antibody that can be made to the transition state should stimulate the reaction of the substrate by making it resemble that transition structure. The problem is that the force constants required for distortion of a molecule are usually very large, and proteins, even enzymes, have only a limited ability to bring about geometric distortion. They can, however, cause ground state destabilization of charged groups, and, if the transition state involves the removal of charge or a decrease in charge density, an antibody to the corresponding uncharged molecule may catalyse a large acceleration in the rate of the reaction.

The most obvious thing that a catalytic antibody should do is bring two molecules together into exactly the correct position to react. Potentially, this induced approximation, if it is sufficiently exact and provides a precise fit, can bring about rate acceleration up to the limit of about 10^8 for 1 M solutions. Antibodies may effect chemical catalysis more directly using acidic and basic groups, if one can stimulate the formation of such groups in the antibodies.

An important question is how much of an advantage one can gain from catalysis by an antibody. The first attempts did not produce very large rate accelerations, but they did demonstrate the feasibility of achieving rate increases with antibodies. We need to learn how to improve these accelerations, to establish which factors cause the acceleration, and to determine the magnitude of the catalysis that can be assigned to each factor. Catalytic antibodies can provide this kind of information more readily than other techniques.

The most important role of catalytic antibodies may be to catalyse specific reactions of rather complex molecules. It is often difficult to use ordinary chemical techniques to probe a particular site or reaction of a molecule that has the potential to participate in many different kinds of reaction. If an antibody could be developed that was specific for a part of such a molecule, or for a particular reaction type in a site on the molecule, it could catalyse specific reactions that are important in developing new drugs, antibiotics, and other chemicals.

These are just a few of the questions and problems that I expect will be addressed during this symposium on this exciting new field of catalytic antibodies.

References

Jencks WP 1966 In: Kaplan NO, Kennedy EP (eds) Current aspects of biochemical energetics. Academic Press, New York p 273–298

Jencks WP 1969 Catalysis in chemistry and enzymology. McGraw Hill, New York

Raso V, Stollar BD 1975 The antibody–enzyme analogy. Comparison of enzymes and antibodies specific for phosphopyridoxyl tyrosine. Biochemistry 14:591–599

A catalytic antibody uses a multistep kinetic sequence

Stephen J. Benkovic,* Joseph Adams,* Kim D. Janda† and Richard A. Lerner†

*The Pennsylvania State University, Department of Chemistry, 152 Davey Laboratory, University Park, PA 16802 and †Scripps Clinic and Research Institute, 10666 North Torrey Pines Road, La Jolla, CA 92037, USA

Abstract. Antibody NPN43C9 catalyses the hydrolysis of a *p*-nitrophenyl ester and *p*-nitroanilide. Kinetic investigations show that hydrolysis of the ester involves at least two steps whose relative importance changes with pH. We propose that the reaction proceeds via formation of an acyl intermediate that deacylates through attack by hydroxide. The rate-limiting step changes from product release at high pH to hydroxide ion-mediated hydrolysis at low pH. This multistep pathway shares similarities with that used by serine proteases.

1991 Catalytic antibodies. Wiley, Chichester (Ciba Foundation Symposium 159) p 4–12

A previously isolated antibody, NPN43C9, catalyses the hydrolyses of a *p*-nitrophenyl ester **1** and *p*-nitroanilide **2** (Janda et al 1988) (Fig. 1). The rate acceleration for anilide hydrolysis relative to the spontaneous hydrolysis at pH 9.0 is approximately one millionfold, it was therefore plausible that this particular antibody might use a kinetic sequence that involved active-site general acid-base catalysis. We tested this hypothesis by a series of pre-steady-state and steady-state kinetic experiments (Benkovic et al 1990).

The behaviour of the steady-state Michaelis–Menten parameters, k_{cat}/K_m and k_{cat}, as a function of pH is exhibited in Fig. 2. Catalytic activity of the antibody increases with increasing pH, controlled by an apparent pK_a of about 9. However, the binding of **2** as a competitive inhibitor of the hydrolysis of **1** is independent of pH ($K_i \approx 260 \pm 80\,\mu$M, pH 7.0; $K_i \approx 190 \pm 60\,\mu$M, pH 9.8). This suggests that substrate and inhibitor binding are not controlled by the ionization state of an active site group (Cleland 1977), and that the pK_a appearing in k_{cat}/K_m as a function of pH has to be explained in another way.

Measurement of the rates of product release from the antibody by stopped-flow competition experiments monitoring the increase in antibody fluorescence (340 nm) on dissociation of the product molecule revealed that the rate constants for product release are in the order *p*-nitroaniline > parent acid **3** > *p*-nitrophenol.

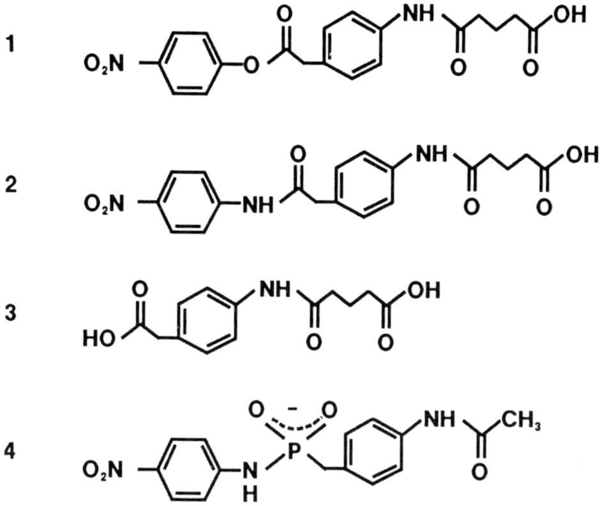

FIG. 1. Structures of substrates and transition state analogues 1–4. Copyright 1990 by the AAAS.

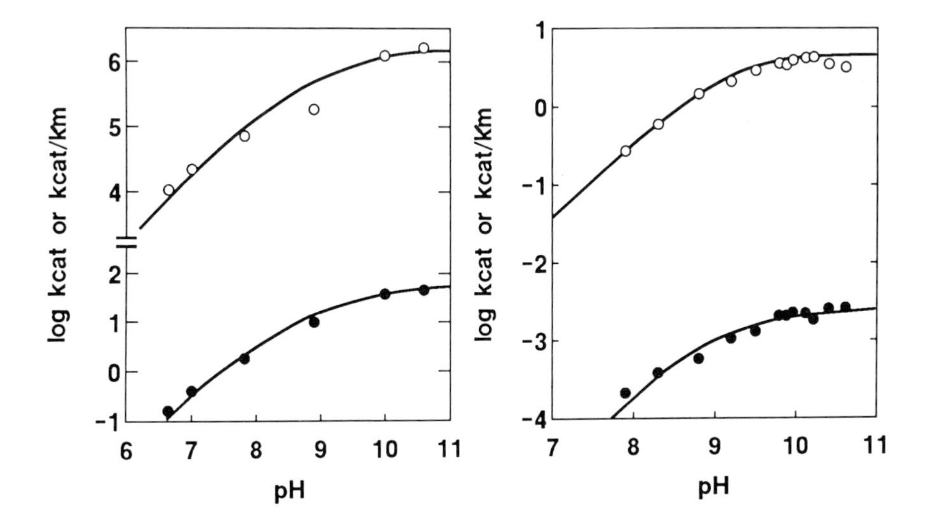

FIG. 2. The pH-dependent steady-state kinetics for hydrolysis of *p*-nitrophenyl (left hand panel) and the *p*-nitroanilide **2** (right hand panel). The calculated lines fitted to $pK_a = 9.1 \pm 0.1$ (k_{cat}, ●); $pK_a = 8.9 \pm 0.1$ (k_{cat}/K_m, ○) for the ester reaction. Calculated lines fitted to $pK_a = 9.0 \pm 0.1$ (k_{cat}, ●); $pK_a = 9.1 \pm 0.1$ (k_{cat}/K_m, ○) for the amide reaction. Copyright 1990 by the AAAS.

Moreover, the rate constant for *p*-nitrophenol release ($45 \pm 10\, s^{-1}$) is, within experimental uncertainty, equal to the k_{cat} observed for ester hydrolysis ($40 \pm 6\, s^{-1}$, pH>9 in Fig. 2). This observation requires that the antibody-catalysed hydrolysis of **1** proceeds through at least two steps whose relative importance changes with pH. The rate of *p*-nitrophenol release as measured by the competition experiments is also independent of pH over the range shown in Fig. 2.

At pH 6.0, the rate of *p*-nitrophenol association with NPN43C9 can be obtained by monitoring the *increase* in absorbance at 404 nm that arises from the partial conversion of *p*-nitrophenol to the phenolate as the binary complex is formed (Fig. 3). The amplitude of the binding curve (in this case 0.9 ± 0.1) reflects the equivalents of *p*-nitrophenol bound at saturation per equivalent of antibody. Since the quantity is half that expected from the total occupancy of both sites, it appears that the pK_a of *p*-nitrophenol has been reduced from about 7.1 to about 6.0. The decrease in the relative amplitude of the absorbance increase (insert, Fig. 3) when the antibody reacts with *p*-nitrophenol in the presence of the phosphonamide transition state analogue, **4**, which was used to induce NPN43C9, showed that they compete for a common binding site. Extrapolation to high ratios of [**4**]/[NPN43C9] implies a 2:1 *p*-nitrophenol: antibody stoichiometry that reflects the homogeneity of the protein preparation and eliminates the possibility of a contaminating esterase.

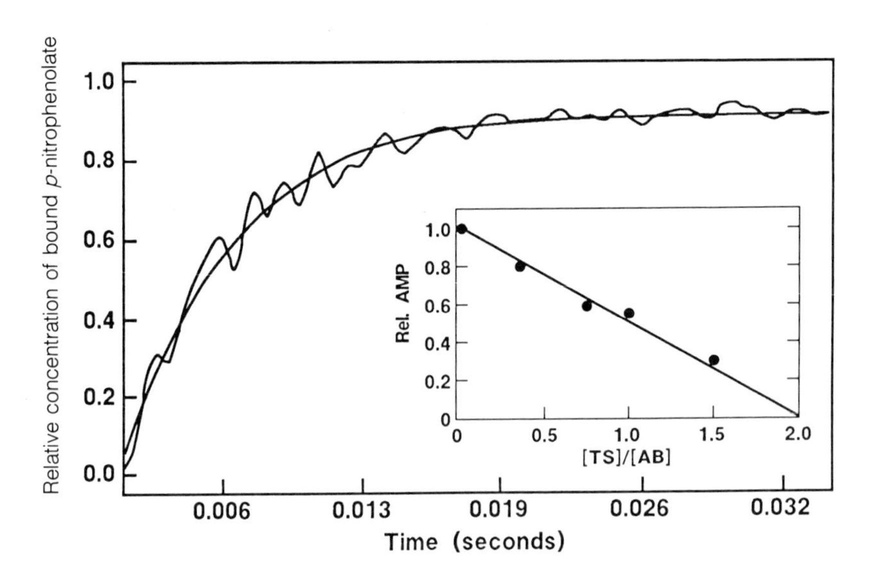

FIG. 3. Time-dependent increase in the amount of *p*-nitrophenolate bound to the antibody as measured in the stopped-flow apparatus. Relative amplitude for the binding of *p*-nitrophenol and antibody as a function of the ratio of the transition state analogue to antibody is shown in the insert. Copyright 1990 by the AAAS.

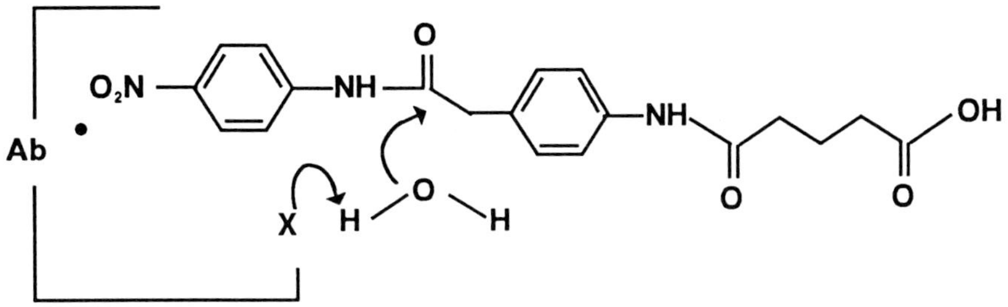

$$\text{but} \quad \frac{k_{cat}\ H_2O}{k_{cat}\ D_2O} \cong 1.0$$

FIG. 4. General base catalysis by a dissociable group at the binding site of antibody NPN43C9. Copyright 1990 by the AAAS.

A simple mechanism featuring general base catalysis of substrate hydrolysis (Fig. 4) may explain the pH profiles in Fig. 1. This pathway invokes a dissociable group at the antibody binding site that removes a proton from an intervening water molecule or from a tetrahedral species (not shown) to catalyse the reaction. Measurement of the pH versus k_{cat} and k_{cat}/K_m profiles for the hydrolysis of **2** in deuterium oxide (data not shown) revealed no evidence for a deuterium solvent isotope effect characteristic of this type of mechanism (Bruice & Benkovic 1966).

In the absence of results supporting an active site group ionizing in the pH range of our observations, we have chosen to interpret the available kinetic evidence in terms of the following kinetic sequence:

$$\mathrm{Ab} + \mathrm{S} \underset{k_{-1}}{\overset{k_1}{\rightleftharpoons}} \mathrm{Ab \cdot S} \underset{k_{-2}}{\overset{k_2}{\rightleftharpoons}} \mathrm{Ab \cdot I} \xrightarrow{k_3\mathrm{OH}^-} \mathrm{Ab \cdot P_1 \cdot P_2}$$

$$\overset{k_4}{\rightleftharpoons} \mathrm{Ab \cdot P_2} \overset{k_5}{\rightleftharpoons} \mathrm{Ab} \qquad (\text{releasing } \mathrm{P_1}, \mathrm{P_2})$$

This sequence features formation of a putative acyl intermediate which deacylates through attack by hydroxide. Using steady-state assumptions, one can solve for k_{cat} and k_{cat}/K_m as a function of pH:

$$k_{cat} = \frac{k_2 k_3 k_4 k_5 \mathrm{OH}^-}{k_3\mathrm{OH}^-(k_4 k_5 + k_2(k_4 + k_5)) + k_4 k_5 (k_2 + k_{-2})}$$

$$k_{cat}/K_m = \frac{k_1 k_2 k_3 \mathrm{OH}^-}{k_3\mathrm{OH}^-(k_2 + k_{-1}) + k_{-1} k_{-2}}$$

At low pH the equations reduce to:

$$k_{cat}(\mathrm{pH}<9.0)=\frac{k_2k_3\mathrm{OH}^-}{k_2+k_{-2}}$$

$$k_{cat}/K_m\ (\mathrm{pH}<9.0)=\frac{k_1k_2k_3\mathrm{OH}^-}{k_{-1}k_{-2}}$$

At high pH the equations reduce to:

$$k_{cat}(\mathrm{pH}>9.0)=\frac{k_2k_4k_5}{k_4k_5+k_2(k_4+k_5)}$$

$$k_{cat}/K_m(\mathrm{pH}>9.0)=\frac{k_1k_2}{k_2+k_{-1}}$$

The pH rate profile for k_{cat} for **1** is thus generated from a change in rate-limiting step from *p*-nitrophenolate release at high pH to hydroxide ion-mediated hydrolysis of Ab·I at low pH. For k_{cat}/K_m the profile probably describes a change from rate-limiting hydroxide ion-mediated deacylation of Ab·I at low pH to diffusional control of antibody–**1** association at high pH. These interpretations agree with the independent measurement of *p*-nitrophenol release and a k_{cat}/K_m value approaching that for diffusional control. For the anilide **2**, both profiles represent a change from rate-limiting hydroxide-mediated hydrolysis of Ab·I at low pH to its acylation at high pH.

We still need direct evidence for the presence of an acyl·antibody intermediate. Given that loss of *p*-nitrophenol or *p*-nitroaniline from the antibody does not occur in the acylation step that forms Ab·I, it is probable that such a species will not accumulate because of the back reaction. Nevertheless, the multistep pathways utilized by NPN43C9 and serine proteases bear a general resemblance. Two marked deviations are the hydroxide-mediated deacylation used by the antibody in contrast to the general acid-base-catalysed deacylation in the proteases (Bender & Kezdy 1964), and the slow *p*-nitrophenol release seen with the antibody.

Despite its limitations, NPN43C9 is a potent catalyst, with k_{cat} values at pH > 9.0 that are within a factor of twenty-five of that of chymotrypsin at pH 7.0 (Zerner et al 1964, Inagami et al 1969). Since product release is the rate-limiting step in antibody-catalysed hydrolysis of the ester, catalytic efficiency may be improved by making discrete structural alterations in the substrate.

References

Bender ML, Kezdy FJ 1964 The current status of the chymotrypsin mechanism. J Am Chem Soc 86:3704–3714

Benkovic SJ, Adams JA, Borders CL Jr, Janda KD, Lerner RA 1990 The enzymic nature of antibody catalysis: development of multistep kinetic processing. Science (Wash DC) 250:1135–1139

Bruice TC, Benkovic SJ 1966 In: Bioorganic mechanisms, vol 1. WA Benjamin, New York, chapter 1

Cleland WW 1977 Determining the chemical mechanisms of enzyme-catalysed reactions by kinetic studies. Adv Enzymol Relat Areas Mol Biol 45:273–387

Inagami T, Patchornik A, York SS 1969 Participation of an acidic group in chymotrypsin catalysis. J Biochem 65:809–819

Janda KD, Schloeder D, Benkovic SJ, Lerner RA 1988 Induction of an antibody that catalyzes the hydrolysis of an amide bond. Science (Wash DC) 241:1188–1191

Zerner B, Bond RPM, Bender ML 1964 Kinetic evidence for the formation of acyl–enzyme intermediates in the chymotrypsin catalyzed hydrolysis of specific chymotrypsin substrates. J Am Chem Soc 86:3674–3679

DISCUSSION

Hilvert: Steve, in what ways have you looked for formation of an acyl-enzyme intermediate?

Benkovic: We looked for its accumulation by rapid-quench experiments and stopped-flow experiments but found none, so there's no direct kinetic evidence. The rate-limiting step in the ester hydrolysis is loss of the product, phenol. The need to rationalize both ester and anilide hydrolysis in terms of a common mechanism favours a mechanism featuring a change in the rate-limiting step. The absence of any pH dependence in studies which involve substrate, inhibitor and product binding over a large pH range suggests that any active site functional groups may have ionized outside the pH span of our experiments. Thus, we have no data that suggest there is a participating group that gives rise to the pK_a of 9.0 seen in the pH rate profiles.

Future experiments using substrates where the leaving group is lost more rapidly may furnish evidence for accumulation of an intermediate. Diethylpyrocarbonate is the only classical chemical reagent that modifies this particular antibody. The amino acid sequence of this antibody also shows that it is somewhat unusual; it has two histidines in what appears to be the binding pocket. Our suspicion is that one of the histidines may participate in substrate hydrolysis. It may have the right chemical characteristics to explain the turnover rates.

Hilvert: Classical partitioning experiments of the acyl intermediate arising from structurally analogous amides and esters wouldn't be applicable to your system because of the participation of the spectator anilide or phenol.

Benkovic: Right, the problem is confounded by not having lost the initial products of the ester or anilide cleavage. There will not be the tell-tale build-up of an intermediate.

Jencks: It's interesting that one of the characteristics you have seen is so close to what has often been found with enzymes—that the chemistry of the catalysis is not rate limiting under many conditions. This indicates that your antibody is acting as a real enzyme.

Benkovic: The acylation is certainly not rate limiting for ester hydrolysis, it is very rapid. It is deacylation that the antibody has trouble with, and release of the product. Many enzymes have trouble with release, although for the antibody this problem is more severe.

Page: Have you looked at any derivatives other than the *p*-nitro compounds?

Benkovic: Yes. We looked at the *p*-chloro compound and found that the pK_a had shifted to a higher pH, as we hoped it would, reflecting the lower reactivity of the chloro compound.

Page: Have you been able to determine the k_{cat} and K_m at the pH plateau and in the pH-dependent region?

Benkovic: Those ratios for the chloro compound are different, for both the anilide and the ester. We have also looked at the *p*-acetoxy compound made by Kim Janda, which is a substrate for this antibody. We have another compound to test with a *p*-formyl group. We hope to determine a full structure/reactivity correlation. Then, by having the appropriate pK_a changes, we should be able to determine more precisely the reactivity of the intermediate.

Green: Do you observe repeated turnovers with the ester, amide and *p*-chloro substrates in your antibody-catalysed reactions?

Benkovic: For the *p*-chloro and *p*-nitro compounds, we have seen thousands of turnovers. We have looked carefully for misacylation producing a suicide inhibitor; we have not found that with the *p*-nitro or with the *p*-chloro compound. I expected to find more of such behaviour as we changed the nature of the *p*-substituent, so it would look less like the initial inducing antigen. As yet, we have not found such inhibition. This antibody, NPN43C9, works very well; it is very stable, even at high pH. You can dialyse out the *p*-nitrophenol, which becomes a serious problem after a while, and the antibody is still active.

Suckling: Steve, the modifications of substrates you are talking about are all in the leaving group. In the original antigen, the butanoate side chain was the link to the protein. Have you tested any modified substrates in which the butanoate was changed?

Benkovic: We have changed some structure on the acid side, without any success.

Janda: The butanoic linker is usually put on just to increase solubility. We have also used simply *N*-acetyl. My group has looked at a series of phosphonate/phosphonamidate haptens and their respective substrates where the phenol acetic acid *N*-acetyl moiety is found in each hapten and substrate. There is no consistent difference in the rates of reaction between the *N*-acetyl and the butanoic acid. Sometimes the *N*-acetyl is a little better.

The *p*-nitro group is essential. We've tried a number of generic amino acids from Sigma; none of those work. The *m*-nitro and the *o*-nitro compounds are not substrates. The cyclohexyl *p*-nitro analogues of the ester and the anilide are not substrates. The saturated ester is an inhibitor with a K_i of about 800 mM.

Jencks: How do you account for the lack of ^{18}O exchange geometrically?

Benkovic: The ^{18}O exchange occurs in the spontaneous reaction, as posited by Bender & Thomas (1961). We saw none in the antibody reaction, which led us to investigate the acyl species.

Paul: Did you try a polyclonal preparation raised to the same antigen?

Benkovic: The polyclonal mixture for that antigen did not show any appreciable catalysis against the anilide.

Page: Steve, in the original generation of the antibody, four-coordinate tetrahedral phosphorus was used as an antigen with the expectation that the antibody would be complementary to this. But in antibody NPN43C9 one of the ligands of the tetrahedral structure has formally become a nucleophile. How do you think the transition state analogue you used generated a nucleophile on the antibody?

Benkovic: How one can link the information content in the transition state analogue to the pathway of substrate turnover used by this particular antibody is something we should discuss. We were looking for simple binding of substrate followed by general acid-base catalysis to give us product. That's not happening. Our result may reflect the diversity of the system. Given all the possible ways of binding this transition state analogue, there may be one in which histidine is in the correct position relative to the substrate to act in that particular pathway.

Page: Could you justify moving away from the tetrahedral transition state stabilization variants? If you had a suitable three-coordinate system, you might generate a complementary nucleophile on the catalytic antibody which might attack the electron-deficient centre.

Benkovic: It's possible. We are looking at other materials we could immunize with to generate antibodies that might achieve catalysis by different pathways.

Green: Have you looked at any other antibodies raised against this hapten?

Janda: Initially, we screened 43 monoclonals by putting them in culture wells, adding substrate, leaving them at 37 °C and looking for development of yellow colour (*p*-nitroaniline). Two monoclonals gave a positive signal: one was much better than the other, so we used the stronger one, NPN43C9, for most of the studies. The *p*-nitrophenol substrate was made by accident; 38 of the 43 monoclonals we tested catalysed the hydrolysis of this to different degrees. Antibody 43C9, the initial clone, which cleaved the amide, was very efficient.

Benkovic: We don't have sufficient information on other antibodies to propose a mechanism. We have a lipase-like antibody that hydrolyses an alkyl ester with a pH rate profile that suggests a single dissociable group. Again that was a very superior antibody in terms of substrate hydrolysis relative to background. We are using our screens now to eliminate antibodies that have rates of 10^3 or less over background.

Schwabacher: It may be useful to screen at an early stage for activity rather than for binding. There's no obvious reason why this particular antibody, which is so interesting, should bind the transition state analogue very strongly, as it is not a good analogue for the mechanism you propose. You may have thrown away some antibodies with higher catalytic activity because they did not bind tightly enough to the hapten in your first screen.

Benkovic: That is a major reason for cloning the system into *E. coli*. Richard Lerner and I felt that we were throwing a great deal of information away. Our hope for the λ technology is that with libraries of about three million Fab fragments we will be able to look for unusual catalysts that may appear only once or twice. The problem now of course is to find adequate ways of screening these numbers.

Shokat: In the studies on the phosphonamidate antibodies, do you know why inhibition by salt was a problem initially?

Benkovic: No, the *m*-nitro compound was a competitive inhibitor in the classic sense. The salt problem with this antibody is still there and we work at a certain salt level in all cases.

Janda: We've only looked at sodium chloride. We don't know if other salts cause this problem.

Hilvert: It is probably the chloride; chloride inhibits many proteolytic enzymes. Presumably, there's an anionic binding site on this protein.

Benkovic: The competitive binding studies did elucidate some of the questions about multiple binding. We can load with excess transition state analogue and see non-competitive effects.

Martin: We also saw binding of multiple molecules of product to the antibody that I will describe in my paper (Martin et al, this volume). During hydrolysis of phenyl acetate, three molecules of phenol bind to the antibody, each with a different effect on the rate of reaction. It is conceivable that more phenols bind but we don't detect them because they have no effect on the kinetics.

References

Bender ML, Thomas RJ 1961 The concurrent alkaline hydrolysis and isotopic oxygen exchange of a series of *p*-substituted anilides. J Am Chem Soc 83:4183–4189

Martin MT, Schantz AR, Schultz PG, Rees AR 1991 Characterization of the mechanism of action of a catalytic antibody. In: Catalytic antibodies. Wiley, Chichester (Ciba Found Symp 159) p 188–200

Structural aspects of antibodies and antibody–antigen complexes

Ian A. Wilson, Robyn L. Stanfield, James M. Rini, Jairo H. Arevalo, Ursula Schulze-Gahmen, Daved H. Fremont and Enrico A. Stura

Department of Molecular Biology, Research Institute of Scripps Clinic, 10666 North Torrey Pines Road, La Jolla, CA 92037, USA

Abstract. The structures of several Fab fragments and Fab–antigen complexes have now been solved at high resolution. These structures of antibodies in complex with proteins, peptides and various other haptens have enabled us to gain insights into the structural basis of immune recognition. Early structures of Fab fragments with and without bound haptens showed the antibody combining sites to be pockets or grooves. More recent Fab–protein complex structures have shown the antibody–antigen interactions to be more extensive with flatter, more undulating binding surfaces. We have solved the structures of three Fab fragments in their native form and as complexes with their respective antigens. Two of these are anti-peptide Fab fragments, the other an anti-progesterone Fab. Comparison of the free and bound structures indicates small but significant changes in the antibody on ligand binding. An analysis of the Fab complexes solved so far indicates that the antibodies can have very differently shaped binding sites, depending on the antigen.

1991 Catalytic antibodies. Wiley, Chichester (Ciba Foundation Symposium 159) p 13–39

Recent important developments have grown out of the ability not only to elicit an immune response to a variety of synthetic immunogens but also to manipulate and engineer antibody combining sites with novel properties. For example, the production of specifically designed antibodies such as catalytic antibodies can now help in the study of enzyme catalysis as well as test our ability to exploit the immune system for the manufacture of biological, medical and biotechnological reagents.

To design experimentally antibodies with novel specificities, we need to understand the structural basis of antibody–antigen interactions. X-ray crystallography has been the primary method through which we have derived detailed information on the structure of antibodies and antibody–antigen complexes. The crystal structures of Fab fragments derived from multiple myeloma immunoglobulins in the 1970s provided our first view of antibody structure. The structure determinations of Fab complexes with small haptens, such as McPC603 with phosphocholine and New with vitamin K_1OH, gave the first indication of how antibodies interact with antigens (see reviews by Amzel & Poljak 1979, Davies & Metzger 1983). Only recently have the structures of antibody–antigen complexes of known specificity been determined. The structures of three specific monoclonal Fab complexes with lysozyme and two with neuraminidase have shown the interacting surfaces to be flatter and more extensive than the clefts or grooves seen with smaller haptens (see review by Davies et al 1990). We have recently solved the structures of three Fab fragments and their respective complexes: an anti-progesterone Fab′ DB3 (J. H. Arevalo, M. J. Taussig & I. A. Wilson, unpublished work 1990), an anti-myohaemerythrin peptide Fab′ B13I2 (Stanfield et al 1990), and an anti-haemagglutinin peptide Fab 17/9 (J. M. Rini, U. Schulze-Gahmen & I. A. Wilson, unpublished work 1990). Comparison of the bound and unbound Fab molecules has enabled us to answer the critical question of whether the antibody changes conformation on antigen binding, and to add to our knowledge of antibody–antigen interactions.

In this paper we will discuss the published structures of Fab fragments and their antigen complexes as well as those determined in our own laboratory. Such analyses of the features of antibody combining sites are essential for understanding the diversity of the immune response and hence immune recognition. These studies will be invaluable for the elucidation of the mechanism of action of catalytic antibodies as well as for the future design of new antibody catalysts.

General structural features of Fab fragments and Fab–antigen complexes

Antibodies can recognize very small haptens, large macromolecules such as proteins, and even larger assemblies such as viruses; however, the antibody molecule remains relatively constant in structure. The two important antigen-binding domains, V_H and V_L, come together to form a β-barrel-like structure which supports the six hypervariable loops or complementarity-determining regions (CDRs) (Fig. 1). It is basically at the open end of this barrel structure that antigens bind through specific interactions with the CDRs. Antibody specificity for a given antigen is determined not only by the tremendous sequence variability in the CDRs but also by the length of the CDRs.

The structures of Fab fragments McPC603, New, Kol and J539 (see reviews by Amzel & Poljak 1979, Davies & Metzger 1983) showed the antibody combining sites to be pockets, clefts or grooves. Study of these Fab fragments complexed with their antigens addressed various questions about the antibody–antigen interaction, such as the nature and extent of antibody–antigen complementarity and the role of charge neutralization. For the phosphocholine–McPC603 complex, charge neutralization and complementarity of fit were shown to be important. Thus, we had a view of the recognition process in which the antigen was surrounded by the antibody in a cleft or pocket.

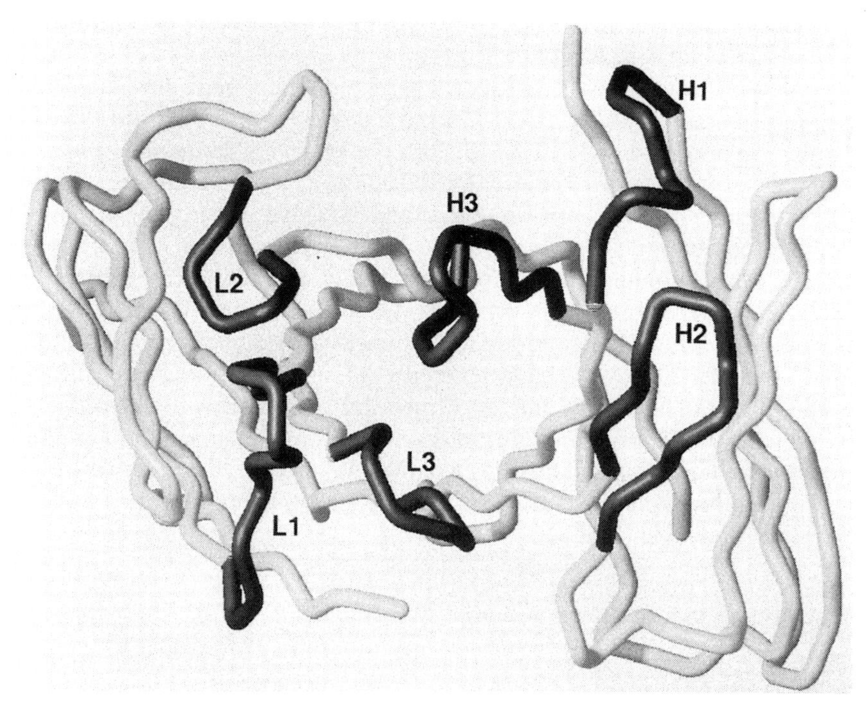

FIG. 1. Anatomy of the antibody combining site. A view down the β-barrel of the Fab variable domain. The association of the V_H and V_L domains produces a β-barrel-type structure. The six hypervariable or complementarity-determining regions are supported by this β-barrel framework. The hypervariable loops (L1–3, H1–3) are shown in dark grey and the framework is shown in light grey. H refers to heavy chain, L to the light chain. The structure of the anti-peptide antibody B13I2 is depicted here. The figure was generated with the program MCS (Connolly 1983).

This view was to change dramatically with the solution of complexes between lysozyme and three different Fab molecules—D1.3 (Amit et al 1986), HyHEL-5 (Sheriff et al 1987a), HyHEL-10 (Padlan et al 1989), and between neuraminidase and Fab molecules NC41 (Colman et al 1987) and NC10 (Colman et al 1989). The antibody did not envelop these larger antigens; the Fab–protein antigen interface was much more extensive than that seen for small haptens and substantially flatter. These five Fab–protein complexes showed the excellent shape complementarity, the role of charge neutralization, and the discontinuous nature of the protein determinants, and revealed that the antigen may undergo small (1–2 Å) but significant changes in its structure on antibody binding (see reviews by Colman 1988, Davies et al 1990). For example, in the D1.3 complex, the complementarity of fit is illustrated by the inability of lysozymes, which have a histidine rather than a glutamine at position 121 to bind to D1.3 (Amit et al 1986). The extent to which charge complementarity is important is highlighted in the HyHEL-5 complex in which two arginines, Arg-45 and Arg-68, on the surface of lysozyme are neutralized by Glu-35 and Glu-50 of the V_H chain (Sheriff et al 1987a). Contrary to the previous commonly held ideas about antibody structure, the HyHEL-5 Fab has a large protrusion in which two side chains, Tyr-33 and Tyr-53 of V_H fit into the lysozyme active site so that the antibody is partially surrounded by the antigen. Thus, from the structure determinations of these protein–Fab complexes, we can no longer describe antibody combining sites simply as pockets into which the antigen will fit. Clearly, when a globular macromolecule binds to an antibody the extent of the interaction depends on the size, amino acid composition and relative disposition of the CDRs. So far, the antibody surface areas buried by the protein antigens range from 690 $Å^2$ to 886 $Å^2$ (Davies et al 1990) and therefore are considerably smaller than the original estimates of the total surface area ($\approx$ 2000–2500 $Å^2$) available for antigen binding. For interaction with 'regularly shaped' globular proteins, the buried surface areas seen in the Fab–protein complexes determined so far probably represent a more realistic upper estimate of the actual surface area used in antigen binding.

Structural studies of free and bound Fab fragments

One critical question that remained was whether there is a change in the antibody structure on antigen binding. In other words, is there a 'lock-and-key' or an induced-fit type of interaction? To address this question we have embarked on an extensive programme of investigating free and bound Fab structures. Because of the reported relatively low success rates in crystallizing antibodies, we set up a number of crystallization trials with many different antibodies. Dr Enrico Stura has now crystallized 20 native Fab molecules and over 30 Fab–antigen complexes. Our antibody programme covers four major areas: (1) anti-peptide and anti-protein antibodies, (2) anti-steroid/anti-growth factor antibodies, (3) catalytic antibodies, and (4) the use of antibodies to help crystallize proteins which will not easily crystallize by themselves. In the first two areas, structures of three free anti-peptide Fab fragments, two Fab–peptide complexes, an anti-progesterone Fab′ fragment and three other steroid–Fab′ complexes have been solved to high resolution ($\leqslant$ 3 Å). The questions concerning antibody–antigen interactions which are addressed by these studies are listed in Table 1.

Anti-peptide Fab fragments and Fab–peptide complexes

Anti-myohaemerythrin peptide Fab fragments and peptide-Fab′ complexes

Several monoclonal antibodies have been raised against a synthetic peptide corresponding to the C-helix (residues 69–87) of myohaemerythrin. Three monoclonals (C,I,L) recognize a determinant, amino acid sequence EVVPH,

TABLE 1 Antibody–antigen issues that can be explored using the structure determinations of Fab–peptide, Fab–steroid and other Fab–antigen complexes

1. Anti-peptide antibody recognition
2. Size of a peptide epitope
3. Peptide conformation—defined structures?
   —same as in protein or as in solution?
4. Success of epitope mapping
5. Binding of chemically related antigens
6. Effect of somatic mutation
7. Conformational change in the antibody
8. Flexibility of V_H/V_L domain
9. Binding sites of haptens versus proteins
10. Nature of antibody combining sites

at the amino end of the peptide, while two others (A,F) recognize a determinant, DFLEKI, at the carboxyl end of the immunizing 19-mer peptide (Fieser et al 1987). In aqueous solution this peptide adopts a nascent helical conformation in the region of the DFLEKI epitope (Dyson et al 1988). Our goal is to determine the structures of these two classes of Fab molecules, which bind to different epitopes on the peptide, and hence see how the peptide conformation is affected by binding at these two different locations. The structures of the free and bound Fab′ B13I2 have now been solved to 2.8 Å resolution (Stanfield et al 1990) and the structure of the unliganded Fab′ B13A2 has been solved to 3 Å resolution (D. H. Fremont, E. A. Stura & I. A. Wilson, unpublished work 1990).

The B13I2 peptide structure determination showed the peptide to be bound in a cleft or pocket on the antibody (Fig. 2). Only seven of the 19 peptide residues can be seen in the electron density map, hence these seven residues constitute the linear peptide determinant. These residues, EVVPHKK, agree fairly well with previous epitope mapping (Fieser et al 1987). The conformation of the peptide bound to the Fab′ is very different from its conformation in myohaemerythrin (Fig. 3) or in aqueous solution. The peptide adopts a type-II β-turn conformation (VPHK) when bound to the antibody but has a predominantly helical conformation in myohaemerythrin (Sheriff et al 1987b). This is not altogether surprising, because His-73, which is the most buried residue in the Fab′–peptide complex, is covalently liganded to one of the iron atoms in the myohaemerythrin molecule. Consistent with the structural data are binding studies which show that the antibody binds well to apomyohaemerythrin but poorly, if at all, to native myohaemerythrin (Stanfield et al 1990). In the peptide–Fab′ complex, the peptide buries 540 $Å^2$ of the antibody surface and the complementarity is excellent, although the fit is not exact (460 $Å^2$ antigen buried surface). As observed in other Fab molecules, there are a large number of

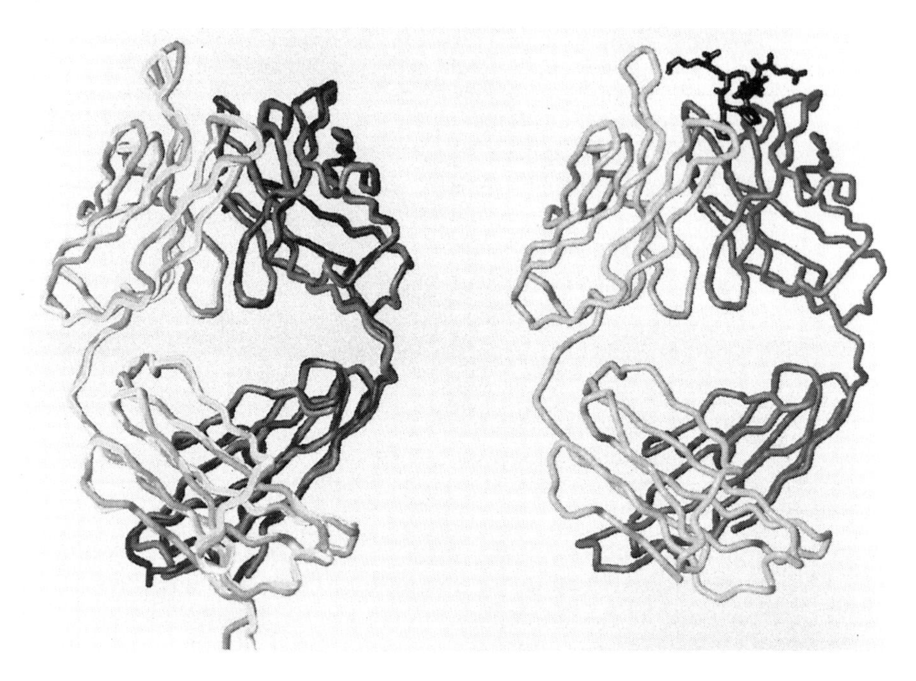

FIG. 2. The structure of the anti-myohaemerythrin peptide Fab′ B13I2 in its native and antigen-bound forms. On the left, the α-carbon traces of the two native Fab molecules in the asymmetric unit of the crystal are shown superimposed; heavy chains on the right, light chains on the left. The variable and constant domains were superimposed separately using the program OVRLAP (Rossman & Argos 1975). On the right-hand side the α-carbon trace of B13I2 in complex with the peptide antigen is shown in the same orientation as the two native Fab fragments.

aromatic residues in the binding site (see reviews by Padlan 1991, Mian et al 1991). Two phenylalanines, Phe-H58 and Phe-H100, sandwich the peptide close to the β-turn. Three tyrosines, L-32, H-56 and H-95, also play a role in the binding site. Sixty-five pairwise van der Waals' contacts are made between the antibody and the peptide as well as 11 specific hydrogen bonds, which include charge neutralization of Lys-75. The majority of these interactions are with the heavy chain, only 12 of the van der Waals' contacts being with the light chain. This distribution of contacts is consistent with 78% of the buried surface on the antibody being contributed by the heavy chain.

The question of whether there are any conformational changes in the Fab molecule when peptide binds can now be answered, because the structures of both the free and the peptide-bound Fab′ B13I2 have been determined (Fig. 2). These were solved independently by molecular replacement from native orthorhombic crystals and hexagonal Fab–peptide complex crystals. In the native crystals, there are two independent copies of the Fab′ structure.

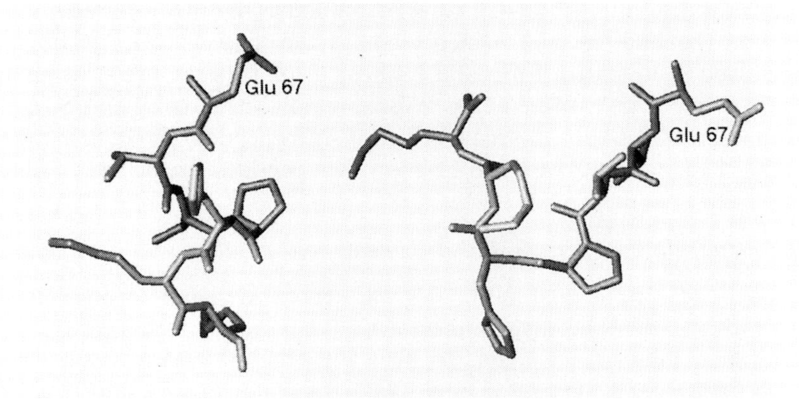

FIG. 3. Comparison of the conformation of a peptide bound to an antibody with its structure in the native protein. The peptide structure on the left corresponds to residues 67–75 of myohaemerythrin. The structure on the right corresponds to the same seven residues when bound to Fab′ B13I2. The first residue of the peptide Glu-67 is labelled. These residues of the peptide in native myohaemerythrin adopt a largely helical conformation but while bound to Fab′ B13I2 are in a type II β-turn conformation. The myohaemerythrin coordinates (Sheriff et al 1987b) were taken from the Brookhaven Protein Data Bank.

This was particularly useful when we were comparing the free and bound Fab molecules to ascertain what changes could be attributed to antigen binding. When superimposed, all three structures overlap well. However, there are variations in the structure of both the native molecules and the complex around residues 129 to 135 of C_H1, a region which is disordered or has high temperature factors in other Fab fragments. There is a structural difference between the free and peptide-bound Fab hypervariable loops, L1 and H3. The deviation in loop L1 is likely to be biologically unimportant, because the variation in the two native structures is as large as either deviation from the complex (Fig. 2); thus the differences in the L1 loop are probably not due to antigen binding. More importantly, there are small but significant changes in loop H3 in the vicinity of residues Pro-99, Phe-100 and Tyr-100B when the peptide binds. The main chain H3 loop in the peptide complex moves about 1.0–1.5 Å from its position in the unbound structure in a segmental type motion (Stanfield et al 1990).

What can we say about the interaction of a peptide with an antibody from this one structure determination? The peptide does have a defined conformation when bound to the Fab fragment and adopts a stable secondary structure commonly observed in proteins. The determinant is linear and consists of seven amino acids. Some of the peptide side chains do not interact closely with the Fab molecule, for example Val-70 interacts only through the main chain hydrogen bonds.

We are now determining the structures of antibodies to other myohaemerythrin peptides. The Fab′ B13A2, which binds to the 'nascent helix' epitope has been solved in its native form and refinement of its structure is in progress (D. H. Fremont, E. A. Stura & I. A. Wilson, unpublished work 1991). Structures of complexes of this Fab′ molecule with the C-helix peptide, and of complexes of the other Fab molecules with myohaemerythrin peptides, are being determined.

Anti-haemagglutinin peptide Fab fragments and peptide–Fab complexes

A panel of 20 monoclonal antibodies was generated against residues 75–110 of the influenza virus haemagglutinin (HA). The majority of these monoclonals recognized a single determinant, residues 101–106, DVPDYA, of the 36-mer peptide (Wilson et al 1984, Houghten 1985). Several Fab fragments from these antibodies have been crystallized either as free molecules (12CA5) or as complexes (26/9, 21/8, 128D4 and 12CA5). The structure of the 17/9 Fab fragment has been solved for the free molecule and for the complex with the 9-mer peptide determinant corresponding to HA residues 100–108. The Fab fragment recognizes the free peptide in solution ($K_a \approx 10^8 M^{-1}$) and the monomeric HA, as well as the fusion-active, low pH form of HA (White & Wilson 1987). As with Fab′ B13I2, crystals of the complex grow under different conditions than do those of the native peptide. The native crystals are monoclinic, $P2_1$, a = 90.3 Å, b = 82.9 Å, c = 73.4 Å, β = 122.5°. Two different Fab-peptide complex crystal forms have been grown and are triclinic, a = 60.1 Å, b = 67.1 Å, c = 73.2 Å, α = 89.9°, β = 101.8°, γ = 96.5° or monoclinic, $P2_1$, a = 63.5 Å, b = 73.4 Å, c = 62.7 Å, β = 117.1°. The native crystals diffract remarkably well ($\leqslant$ 1.9 Å) and the structure has been solved at 2.0 Å resolution. The triclinic crystals are large thin plates (0.8 × 0.4 × 0.05 mm) and have severe twinning problems; this structure has now been determined at 2.9 Å resolution. The monoclinic complex crystals are even thinner (0.8 × 0.02 × 0.01 mm) and grow as long needles. Because of the shortage of these monoclinic crystals, a complete data set has been collected from just one crystal at −150 °C and the structure of this Fab–peptide complex has been determined at 3.0 Å resolution.

The 9-mer peptide can be clearly traced in the electron density map of both complex crystals. The determinant is again linear and appears to contain seven residues. Differences are apparent in the conformation of some of the CDRs between the free and bound Fab molecules. In particular, the H3 loop is significantly different when peptide is bound. Refinements of structures of two Fab fragments in the asymmetric unit of the native crystals and of those of the three molecules in the asymmetric units of the two Fab–peptide complexes are in progress (J. M. Rini, U. Schulze-Gahmen & I. A. Wilson, unpublished work 1991). This antibody also binds extremely well to the monomeric haemagglutinin 'tops' (HA1 residues 28–328; affinity determination in progress). Efforts are being made to crystallize the HA 'tops' with Fab 17/9.

Anti-progesterone Fab′ fragments and steroid–Fab′ complexes

A panel of high affinity anti-progesterone antibodies ($K_a \approx 10^9\,M^{-1}$) has been raised against 11α-hemisuccinyl progesterone conjugated to bovine serum albumin (BSA) (Wright et al 1982). The Fab′ fragment from one of these antibodies, DB3, has been crystallized, with progesterone and with each of seven related but structurally different steroids (Stura et al 1987). This gives us an excellent opportunity to investigate how a single antibody can recognize a series of chemically related steroids and hence how the antibody combining site can accommodate different antigens. Clearly, these are critical issues to consider in the rationale behind the design and production of catalytic antibodies.

Large single hexagonal crystals have been grown of the native DB3 Fab′ and of Fab′–steroid complexes of progesterone, 11α- and 11β-hydroxyprogesterone, 11α-hemisuccinyl progesterone, 4-iodobenzoyl-11α progesterone, 5α- and 5β-pregnanedione and 3α-aetiocholanolone. All the complexes have been co-crystallized and are isomorphous with crystals of the native Fab′. The crystal morphology varies from long hexagonal rods for the native peptide to spherical and elipsoid shapes, depending on which steroid is bound (Stura et al 1987). In order to demonstrate that steroid was indeed present, we reacted the crystals with 2,4-dinitrophenylhydrazine, which reacts with free steroidal ketone groups, causing a colour change. The complex crystals changed colour as expected, but the native crystals also gave indications of some bound steroid. Further analysis showed that the antibody derived from mouse ascites carried ≈10–15% steroid throughout the Fab′ production and purification (Stura et al 1987). A steroid-free native DB3 has now been produced in cell culture in Dr Michael Taussig's laboratory. The structures of the native Fab′ molecule (2.8 Å) and of its complexes with progesterone (2.4 Å), iodobenzoyl progesterone (2.8 Å) and 11α-hemisuccinyl progesterone (3.0 Å) have been solved but not yet refined. A preliminary analysis shows the progesterone is bound in a cleft or slot in the antibody combining site. As with the anti-peptide Fab fragments there appear to be changes in the Fab′ molecule on antigen binding. In particular, a tryptophan side chain which occupies the binding site in the free Fab′ is in a substantially different location when progesterone is bound (J. H. Arevalo, M. J. Taussig, I. A. Wilson, unpublished work 1991). Also of interest is the lack of density seen for the iodobenzoyl group in the iodobenzoyl progesterone–Fab complex, which indicates either positional disorder or cleavage of the ester linkage during the co-crystallization. This is reminiscent of a reported esterase-like activity of an antibody with steroid esters where the hydrolysis was stoichiometric rather than catalytic (Kohen et al 1980). Thus, we are investigating whether our antibody has esterase activity.

From this study, the stage is now set for the analysis of how an antibody can interact with different antigens. The steroids used, although chemically related, do have different conformations (see Duax & Norton 1975).

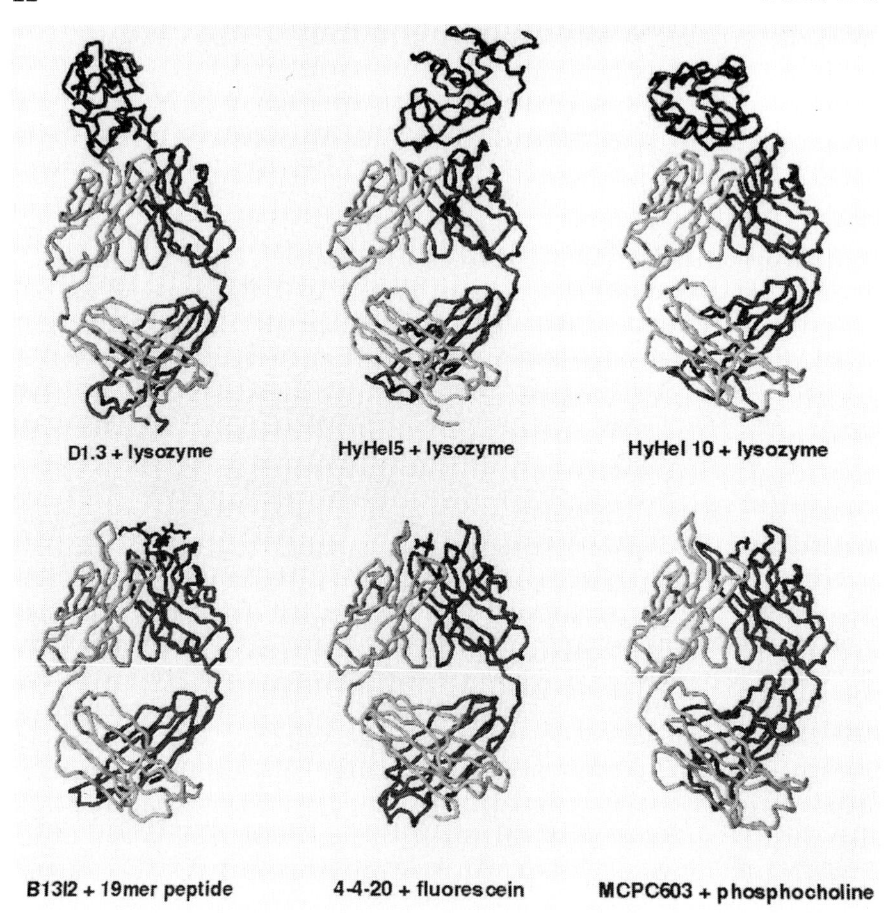

FIG. 4. The structures of Fab–antigen complexes. The α-carbon traces of six Fab–antigen complexes are depicted. The heavy chain of the Fab is depicted in dark grey, the light chain is light grey and the antigen is black. The coordinates for the HyHEL-5 (Sheriff et al 1987a), HyHEL-10 (Padlan et al 1989), 4-4-20 (Herron et al 1989) and McPC603 (Satow et al 1986) complexes were taken from the Brookhaven Protein Data Bank. The coordinates for the B13I2 peptide complex are from our laboratory and the D1.3 lysozyme coordinates were provided by Dr S. E. V. Phillips and Dr R. J. Poljak. The figures were calculated using the program MCS (Connolly 1983) on a Sun-4 Spark work station. (From Wilson et al 1991.)

The A ring can be severely displaced out of the plane of the B, C and D rings, as in the 5β-androstane-3,17-dione-type structures compared to the almost planar 5α-type structures. Another interesting study of immunological importance involves the comparison of the germline and somatically mutated structures of Fab fragment DB3 and should evaluate what important changes, if any, have occurred during the maturation process.

TABLE 2 The buried surface areas accessible to solvent in Fab–antigen complexes

Antibody	*Buried surface area Å²*	*Antigen*	M_r	*Total surface area Å²*	*Buried surface area Å²*	*% buried*
4-4-20	308	Fluorescein	320	282	266	94
DB3	303	Progesterone	314	277	239	86
McPC603	161	Phosphocholine	169	169	137	81
B13I2	540	C-helix peptide	818	701	462	66
F17/9	468	Haemagglutinin peptide	1055	742	436	59
HyHEL-5	746	Lysozyme	14 000	5436	750	14
HyHEL-10	721	Lysozyme	14 000	5414	774	14
D1.3	690	Lysozyme	14 000	5564	680	12
NC41	886	Neuraminidase	50 000	14 648	879	6

The surface area calculations for 4-4-20, McPC603, HyHEL-5, HyHEL-10, D1.3 and NC41 were taken from Davies et al (1990). Buried surface areas for B13I2, DB3 and F17/9 were calculated with the program MS (Connolly 1983) using a 1.7 Å probe radius and van der Waals' radii as in Case & Karplus (1979). These probe and van der Waal's radii are identical to those used in the Davies et al (1990) calculations.

Catalytic antibodies

We are working on the structures of several catalytic antibodies and have obtained crystals of three. Large crystals of one of the first catalytic antibodies, 8D2, which catalyses an esterase reaction (Tramontano et al 1986), have been grown in the presence and absence of transition state analogues. Smaller crystals have been grown of the Diels–Alder antibody 1E-9 and the Claisen rearrangement antibody 27G5 in collaboration with Dr Donald Hilvert (see Hilvert, this volume). Attempts to improve the quality of these crystals is in progress, so that we can determine the structure of a catalytic antibody and elucidate its catalytic mechanism.

The nature of antibody combining sites

The structures of Fab fragments complexed with five proteins, a peptide, phosphocholine, vitamin K_1OH, and fluorescein (Herron et al 1989), have been reported. Unpublished structures solved include the anti-peptide Fab 17/9 and the anti-progesterone Fab described here. Structures of antibodies with other antigens, such as nucleic acids (anti-single-stranded DNA Fab BV04-01, A. B. Edmundson et al, unpublished paper, XVth Congr, Int Union of Crystallography, Bordeaux, July 1990), carbohydrates (Fab Ser-155.4, Rose

TABLE 3 Interaction of the antigen with the hypervariable loops of the antibody

Antibody	*L1*	*L2*	*L3*	*H1*	*H2*	*H3*	V_L(%)	V_H(%)
McPC603	−	−	+	+	+	+	47	53
NEW	+	−	+	−	+	+		
4-4-20	+	−	+	+	+	+	40	60
B13I2	+	−	+	+	+	+	22	78
F17/9	+	−	+	−	+	+	26	74
DB3	+	−	+	+	+	+	35	65
D1.3	+	+	+	+	+	+	43	57
HyHEL-5	+	+	+	+	+	+	41	59
HyHEL-10	+	+	+	+	+	+	43	57
NC41	−	+	+	+	+	+	44	56
	8	4	10	8	10	10		

In the structures determined so far, the antigen interacts with four to six of the complementarity-determining regions (CDRs). In the hapten and peptide complexes, L2 has not yet been observed to interact with the antigen. In the protein–Fab complexes, not all CDRs are necessarily used for antigen binding as is shown for the anti-neuraminidase antibody NC41. L3, H2 and H3 are the loops used most frequently in these complexes. The complexes are: McPC603 with phosphocholine; NEW with vitamin K_1OH; 4-4-20 with fluorescein: B13I2 and 17/9 with peptides; DB3 with steroids; D1.3, HyHEL-5 and HyHEL-10 with lysozyme; and NC41 with neuraminidase. See text for references and reviews of these structures. On the right is shown the relative surface area buried on V_H and V_L by the antigen as a percentage of the total buried surface. The calculation with Fab NEW is not possible as coordinates of the complex are not available. Coordinates of McPC603, HyHEL-5 and HyHEL-10 are taken from the Brookhaven Databank. The 4-4-20 coordinates were kindly provided by Dr Allen Edmundson, those for D1.3 by Drs Simon Phillips and Roberto Poljak, and those for NC41 by Drs Peter Colman and Bill Tulip. The B13I2, 17/9 and DB3 coordinates are from our laboratory.

et al 1990), dinitrophenyl-spin-label (Fab ANO2, Brunger et al 1991) and 2-phenyloxazolone (Fab NQ10/12.5, Alzari et al 1990), are well advanced, as are the structures of many other interesting Fab fragments and Fab complexes.

Past experience has shown that it is always dangerous to generalize about antibody structures, but some basic features are already apparent. The Fab fragment is very flexible; the 'elbow' angle which connects the variable and constant domains varies between 130° and 180° in different Fab structures. The relative disposition of V_H and V_L varies by up to 10° in different antibodies and may also change significantly on antigen binding, as for the Fv and Fab fragments of D1.3 (Bhat et al 1990) and the anti-single-stranded DNA Fab BV04-01 (A. B. Edmundson et al, as above). A representative illustration of the different antibody complexes determined so far is shown in Fig. 4. The surface buried on the antibody by the antigen ranges from 161 Å^2 for phosphocholine to 886 Å^2 for neuraminidase (see also Davies et al 1990). The small antigens, such as fluorescein and progesterone, are essentially buried (89–94%),

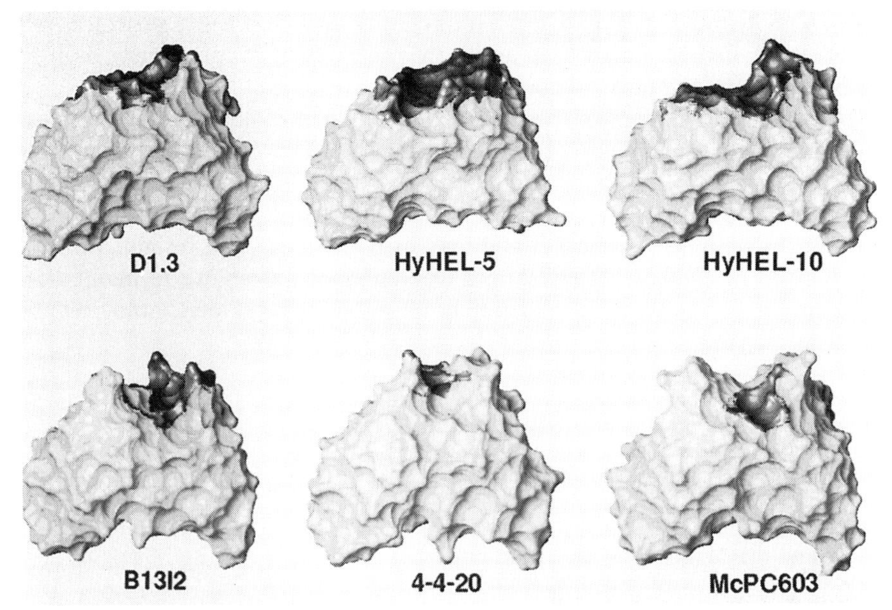

FIG. 5. The different shapes of antibody combining sites. Solid surface representation of the variable domains of the six Fab fragments shown in Fig. 4. These Fab structures were solved complexed with antigen but for clarity the antigen is not shown. The domains are in the same orientations as shown in Fig. 4. The buried surface of the antigen-binding site is highlighted in dark grey. The coordinates used are as described in Fig. 4. The figures were calculated with the program MCS using standard atomic radii and a probe radius of 3.5 Å. The use of such a large probe radius emphasizes all but the major features of the Fab surface giving a smoother representation of the shape of the antigen-binding pocket. (From Wilson et al 1991.)

whereas peptide antigens are 60–65% buried and the proteins obviously less so (Table 2). In the structures already determined, there is more surface area buried on the heavy chain (53–78%) than the light chain (22–47%) (Table 3). However, we have only a limited set of Fab–antigen complexes, so it is too early to make generalizations.

Four to six hypervariable loops are used in binding these antigens. Even with a large antigen such as neuraminidase the NC41 antibody uses only five of its six CDRs (not L1) in its interaction with antigen (Colman et al 1987). In the six hapten and peptide complexes solved so far, the common feature is that L2 is not used for binding (Table 3). These antibodies generally have much longer L1 loops than the corresponding anti-protein antibodies (see Fig. 4).

Finally, it is clear that there is substantial variation in the shape of the antibody combining sites, depending on the ligand (Fig. 5). The structures of antibodies to smaller antigens such as peptides and haptens show that the combining sites are clearly pockets, clefts or slots. The interacting surfaces of the anti-protein

Fabs tend to be flatter and more undulating and may even protrude (see HyHEL-10; Fig. 5), rather than adopt the concave shapes previously thought to be a feature of antibody combining sites.

In conclusion, there has been an explosion in the numbers of antibody structures and their complexes that are being solved. The analysis of their binding sites can now be used to understand better the way in which antibodies interact with antigens. Such structural data will be invaluable in the production of specifically designed antibodies and may provide major new insights into the mechanism of catalysis. They should also enhance our ability to manipulate the immune system for the production of novel biological, biotechnological and medical reagents.

Acknowledgements

We would like to thank our collaborators, Dr Richard Lerner for the production of the original clones of the anti-peptide antibodies B13I2, B13A2 and 17/9, Dr Michael Taussig, AFRC, Babraham for the anti-progesterone Fab′ DB3, and Dr Don Hilvert for the Diels–Alder and the Claisen rearrangement antibodies. Gail Fieser and Rod Samodal gave excellent technical assistance in the antibody and Fab production. We thank Chris Hassig for his help with Fig. 4. This work was supported in part by National Institutes of Health grants AI-23498, GM-38794 and GM-38419 (to I.A.W.) and NIH training fellowship AI-07244 (R.L.S.). Dr James M. Rini is supported by a Centennial Fellowship from the MRC (Canada) and Dr Ursula Schulze-Gahmen by a DFG Fellowship (FRG). Daved H. Fremont is a graduate student in the Chemistry Department at the University of California, San Diego.

References

Alzari PM, Spinelli S, Mariuzza RA et al 1990 Three-dimensional structure determination of an anti-2-phenyloxazolone antibody: the role of somatic mutation and heavy/light chain pairing in the maturation of an immune response. EMBO (Eur Mol Biol Organ) J 9:3807–3814

Amit AG, Mariuzza RA, Phillips SEV, Poljak RJ 1986 Three-dimensional structure of an antigen–antibody complex at 2.8 Å resolution. Science (Wash DC) 233:747–753

Amzel LM, Poljak RJ 1979 Three-dimensional structure of immunoglobulins. Annu Rev Biochem 48:961–997

Bhat TN, Bentley GA, Fischmann TO, Boulot G, Poljak RJ 1990 Small rearrangements in structures of Fv and Fab fragments of antibody D1.3 on antigen binding. Nature (Lond) 347:483–485

Brunger AT, Leahy DJ, Fox RO 1991 The 2.9 Å resolution structure of an anti-dinitrophenyl-spin-label monoclonal antibody Fab fragment with bound hapten. J Mol Biol, in press

Case, DA, Karplus M 1979 Dynamics of ligand binding to heme proteins. J Mol Biol 132:343–368

Colman PM 1988 Structure of antibody–antigen complexes: implications for immune recognition. Adv Immunol 43:99–132

Colman PM, Laver WG, Varghese JN et al 1987 Three-dimensional structure of a complex of antibody with influenza virus neuraminidase. Nature (Lond) 326:358–362

Colman PM, Tulip WR, Varghese JN et al 1989 Three-dimensional structures of influenza virus neuraminidase–antibody complexes. Philos Trans R Soc Lond B Biol Sci 323:511–518

Connolly ML 1983 Analytical molecular surface calculations. J Appl Crystallogr 16:548–558

Davies DR, Metzger H 1983 Structural basis of antibody function. Annu Rev Immunol 1:87–117

Davies DR, Padlan EA, Sheriff S 1990 Antibody–antigen complexes. Annu Rev Biochem 59:439–473

Duax WL, Norton DA 1975 Atlas of steroid structure. Plenum Press, New York, vol 1 p 238–240,334–336,414–416

Dyson JH, Rance M, Houghten RA, Wright PE, Lerner RA 1988 Folding of immunogenic peptide fragments of proteins in water solution. II. The nascent helix. J Mol Biol 201:201–217

Fieser TM, Tainer JA, Geysen MH, Houghten RA, Lerner RA 1987 Influence of protein flexibility and peptide conformation on reactivity of monoclonal anti-peptide antibodies with a protein α-helix. Proc Natl Acad Sci USA 84:8568–8572

Herron JN, He X, Mason ML, Voss EW, Edmundson AB 1989 Three-dimensional structure of a fluorescein-Fab complex crystallized in 2-methyl-2,4-pentanediol. Proteins Struct Funct Genet 5:271–280

Hilvert D 1991 Antibody catalysis of carbon–carbon bond formation. In: Catalytic antibodies. Wiley, Chichester (Ciba Found Symp 159) p 174–187

Houghten RA 1985 General method for the rapid solid-phase synthesis of large numbers of peptides: specificity of antigen–antibody interaction at the level of individual amino acids. Proc Natl Acad Sci USA 82:5131–5135

Kohen F, Kim JB, Barnard G, Lindner HR 1980 Antibody-enhanced hydrolysis of steroid esters. Biochim Biophys Acta 629:328–337

Mian IS, Bradwell AR, Olson AJ 1991 Structure, function and properties of antibody binding sites. J Mol Biol 217:133–151

Padlan EA 1990 On the nature of antibody combining sites: unusual structural features that may confer on these sites an enhanced capacity for binding ligands. Proteins Struct Funct Genet 7:112–124

Padlan EA, Silverton EW, Sheriff S, Cohen GH, Smith-Gill SJ, Davies DR 1989 Structure of an antibody–antigen complex: crystal structure of the HyHEL-10 Fab–lysozyme complex. Proc Natl Acad Sci USA 86:5938–5942

Rose DR, Cygler M, To RJ, Przybylska M, Sinnott B, Bundle DR 1990 Preliminary crystal structure analysis of an Fab specific for a *Salmonella* O-polysaccharide antigen. J Mol Biol 215:489–492

Rossmann MG, Argos P 1975 A comparison of the heme binding pocket in globins and cytochrome b_5. J Biol Chem 250:7525–7532

Satow Y, Cohen GH, Padlan EA, Davies DR 1986 Phosphocholine binding immunoglobulin Fab McPC603. An X-ray diffraction study at 2.7 Å. J Mol Biol 190:593–604

Sheriff S, Silverton EW, Padlan EA et al 1987a A three-dimensional structure of an antibody–antigen complex. Proc Natl Acad Sci USA 84:8075–8079

Sheriff S, Hendrickson WA, Smith JL 1987b Structure of myohemerythrin in the azidomet state at 1.7/1.3 Å resolution. J Mol Biol 197:273–296

Stanfield RL, Fieser TM, Lerner RA, Wilson IA 1990 Crystal structures of an antibody to a peptide and its complex with peptide antigen at 2.8 Å. Science (Wash DC) 248:712–719

Stura EA, Arevalo JH, Feinstein A, Heap RB, Taussig MJ, Wilson IA 1987 Analysis of an anti-progesterone antibody: variable crystal morphology of the Fab′ and steroid–Fab′ complexes. Immunology 62:511–521

Tramontano A, Janda KD, Lerner RA 1986 Catalytic antibodies. Science (Wash DC) 234:1566–1570
White JM, Wilson IA 1987 Anti-peptide antibodies detect steps in a protein conformational change: low-pH activation of the influenza virus hemagglutinin. J Cell Biol 105:2887–2896
Wilson IA, Niman HL, Houghten RA, Cherenson AR, Connolly ML, Lerner RA 1984 The structure of an antigenic determinant in a protein. Cell 37:767–778
Wilson IA, Rini JM, Fremont DH, Stura EA 1991 Use of X-ray crystallographic analysis of Fabs and Fab–antigen complexes to investigate the structural basis of immune recognition. Methods Enzymol, in press
Wright LJ, Feinstein A, Heap RB, Saunders JC, Bennett RC, Wang MY 1982 Progesterone monoclonal antibody blocks pregnancy in mice. Nature (Lond) 295:415–417

DISCUSSION

Benkovic: Do you see any intrinsic limitations in using mainly loops to bind ligands, as antibodies do, rather than the structures that enzymes use for catalysis? Is there a structural limitation to what we might achieve by antibody catalysis?

Wilson: I don't see any limitation. There is an enormous range of possible substitutions for the loops plus additional variation in the lengths of the CDR loops. Superficially, the antibody combining site looks like TIM (triose phosphate isomerase), but it has a completely different structure. TIM has a parallel β-barrel conformation surrounded by α-helices, but the substrates are still bound at the end of the barrel. In antibodies, the V_H and V_L anti-parallel β-sheets come together to form a barrel-like structure (Fig. 1). Antigens bind at the end of the barrel to the CDR loops that connect the β-strands of the sheets.

Hilvert: The anti-progesterone antibody's conformational adjustment mimics induced fit in enzymes.

Wilson: In some cases, the loops appear to mimic induced fit. They seem to be able to move in a segmental motion by about 1–2 Å. For the 17/9 Fab fragment, we see a more substantial rearrangement in the CDR H3 loop when the peptide antigen binds (J. M. Rini, U. Schulze-Gahmen, I. A. Wilson, unpublished).

Benkovic: Which enzymes use loops so extensively for recognition of substrate?

Wilson: Enzymes tend not to use loops; often the recognition site is right at the end of barrel-type structures.

Benkovic: Is there any intrinsic advantage of barrels versus loops for catalysis?

Lerner: An evolutionary reason for not having loops in enzymes is that if there were a mutation in a rigid structural motif, the protein would be eliminated. If the recognition site were in a loop, the enzyme might be able to accumulate mutations and change its characteristics.

Wilson: Let us not overgeneralize; enzymes do have loops. TIM has a loop that moves in when either the phosphate or the substrate dihydroxyacetone phosphate binds, as first observed by Phillips et al (1977).

Lerner: What would happen if there were an accidental substitution into that loop?

Wilson: Presumably, if it were a critical binding residue or if the substitution prevented any conformational changes required for binding, the substrate would no longer bind.

Hilvert: In carboxypeptidase A, Tyr-248 undergoes a 12–15 Å shift to cover up the substrate (e.g. Rees & Lipscomb 1981). The backbone conformation of the loop that contains this tyrosine does not change significantly, however. It was originally thought that Tyr-248 was an essential general acid catalyst, since the phenolic hydroxyl group is within hydrogen-bonding distance of the scissile bond, but this seems not to be the case (Hilvert et al 1986).

Benkovic: Dihydrofolate reductase has a loop and we can make up to seven mutations in it with no effect on its catalytic activity. The loop contacts the methotrexate or dihydrofolate at the benzoyl side chain.

Wilson: Antibodies also have, similar to enzymes, domain movements. Allen Edmundson has an anti-single-stranded DNA antibody, BV04-01, in which V_H moves by at least six degrees relative to V_L when a trinucleotide binds (A. B. Edmundson et al, unpublished paper, XVth Congr, Int Union of Crystallography, Bordeaux, July 1990).

Green: Can one make any general comparisons between the antibody–hapten structures and the extraordinarily tightly bound avidin–biotin structure? Could one exploit anything in the avidin structure to design an antibody that would bind its hapten more strongly?

Wilson: Antibodies often have a more limited range of residues at those interfaces. Some antibodies use aromatic residues extensively for antigen binding, particularly tyrosines and tryptophans (see reviews by Padlan 1990, Mian et al 1991). The question is what makes an antibody combining site so good at binding antigens? Significant energy for binding can be obtained because of the large surface areas of these aromatic residues. With only one or two rotatable side chain torsional angles, they also have relatively low conformational entropy. Such features may give the properties required for ligand binding (see reviews by Davies et al 1990, Padlan 1990, Mian et al 1991).

Benkovic: How many water molecules are there in the smallest antibody antigen-binding sites?

Wilson: We've seen some. However, at the resolution of our structures at present, I would not be confident about the exact number. R. J. Poljak's group has identified specific bound water molecules at the antibody–antigen interface (personal communication).

Benkovic: Are there any examples of loops closing over binding sites in antibodies and shutting out water entirely?

Lerner: Not yet.

Green: In the streptavidin structure there is a loop that closes over the biotin (Weber et al 1989).

Lerner: Also in TIM.

Hansen: So the top of the peptide bound to your antibody is completely solvated by water?

Wilson: At the present resolution, we can't be confident about the location of any water molecules, but presumably there are some there.

Paul: Are there substantial repulsive interactions between the antigen and the antibody?

Wilson: Calculations by Novotny et al (1989) suggested that only a small number of residues from either the antibody or the antigen make significant positive contributions to binding, the rest are either negative or neutral.

Paul: Early in the immune response the antibodies bind antigens with a low affinity (10^6 or less). They are also polyreactive; for example, antibodies that bind DNA may bind structurally dissimilar antigens, such as phospholipids and polypeptide antigens (Hartman et al 1989, Logtenberg 1990). This must mean antibodies have a fairly plastic active site that is capable of conformational change. How rigid is the antibody active site in your experience?

Wilson: In all the antibodies we have studied, at least one or two loops change their position during complex formation. They are not completely rigid.

Paul: Are the changes substantial enough to account for the wide range of antigens that can be bound by a low affinity antibody?

Wilson: I don't think we have enough information yet. The calculations are not good enough to tell us what constitutes a substantial change. We don't know if it is 5 Å, 1 Å or just 0.1 Å.

Jencks: What is the disadvantage of having many possible conformational states, compared to the advantage that you get from binding as a result of the conformational changes?

Wilson: I don't think there is much floppiness or conformational flexibility in the CDR loops. These changes could be described as small, rather rigid segmental changes in the backbone of the loop with greater variation in the side chain torsions, as in the B13I2 structure (Stanfield et al 1990). The only time we have seen a loop change more dramatically is in the 17/9 anti-peptide Fab.

Jencks: There is a small number of states?

Wilson: Yes, in the structures we have at present; however, we are still in early days. There is a whole host of complexes being studied and we do not have sufficient information to make general rules yet.

Hansen: Could you say a bit more about the cross-reactivity, or lack thereof, of the myohaemerythrin peptide and protein? How do you see the issue of cross-reactivities between anti-peptide antibodies and the corresponding antibody?

Wilson: For that anti-myohaemerythrin antibody, B13I2, the histidine in the epitope is actually liganded to the iron atom in myohaemerythrin (Sheriff et al 1987). That's why it reacts a lot better with the apomyohaemerythrin. The anti-haemagglutinin antibody peptide, 17/9, seems to bind extremely well to both peptide and haemagglutinin monomer (U. Schulze-Gahmen, I. A. Wilson, unpublished).

Hansen: Can you measure cross-reactivity between the myohaemerythrin metalloprotein and the antibody to the peptide?

Wilson: It's very low compared to apomyohaemerythrin, less than 10^3–10^4 M^{-1} in a competition assay against the 19-mer peptide.

Hansen: A number of us are interested in putting analogues within peptides and hoping for cross-reactivity to the native protein. Would you consider that to be a legitimate strategy?

Wilson: Not every antibody will cross-react, but some will. If you look hard enough you will find some that do.

Lerner: Antibodies are magnificent physical chemists. They have only about six weeks to increase their binding energy by 20 kcal. Ian has said that, basically, they will use any solution that is energetically favourable. If burying the hapten is more important, let's say, than developing some charge, that is exactly what will happen. All antibodies must compete with their neighbours for binding energy, otherwise they drop out of the running. You must continue to stimulate the system for the players to survive and anybody who falls off by a mistaken mutation leaves the field. At the start of an immune response there are a lot of solutions to low energy. As the immune response proceeds there are probably fewer and fewer ideal solutions and any protein that makes a mistake drops out. With antibodies we have the privilege of seeing a screening system that tells us an energetic solution to a problem.

The system is also reading the energetics of the antigen. The system must see the conformation of the antigen within some very narrow ΔG range and it must see it over and over again. Imagine a case in which the primary recognition event was of some very high energy conformer of Ian Wilson's peptide, way out in the Boltzmann distribution. The chances of ever seeing that conformer again are very small. So not only is the immune system doing its best to increase its binding energy by that 20 kcal, but in a curious way it is also obliged to read the Boltzmann distribution.

Jencks: Are you saying that the system responds directly to the affinity of binding to the antigen?

Lerner: Yes. It takes about 10 individual, and in theory isolated, inductive events to get an immune system working automatically. At the same time, the system is making a product which removes the inducer. That competition, almost by definition, leads to the higher binding energy.

Paul: The caveat is that too high a binding energy might eventually result in down-regulation of the receptors, shutting off the lymphocytes that are producing the antibody. So very high binding affinity might be self defeating.

Lerner: I don't think there's any evidence for that whatsoever.

Paul: There is. High affinity antibody-producing B lymphocytes are preferentially tolerized by antigen (Goodnow et al 1989).

Lerner: There is no tolerance in these situations. The experiments being described here are immunizations, often on carrier molecules, in adjuvant. I don't think anybody has been able to produce tolerance in a system like this in adjuvant.

Paul: Antibody-producing lymphocytes may or may not be dependent on adjuvant and carrier effects provided by molecules in their microenvironment. In transgenic mice deficient for lysozyme, tolerant B lymphocytes can be generated in the absence of adjuvant and carrier molecules by the expression of lysozyme (Goodnow et al 1989).

Green: Ian, in your crystallization procedure for obtaining Fab–hapten complexes, do you crystallize the free Fab fragment and then diffuse in the hapten?

Wilson: No, we always co-crystallize. The crystallization conditions are often substantially different for the native antibody than for the complex. The binding site is often blocked or restricted because of crystal packing in the native antibody. Also, changes in the loops may make it difficult to get isomorphous crystals for the native Fab and the Fab–antigen complex.

Burton: From your crystal data one gets the idea that the CDR3 moves out of the way and the hapten enters the antibody combining site. Is it not that the loop is flexible and, when you crystallize, you catch one structure? Then, when you crystallize again, you get another structure with the hapten in place.

Wilson: No, because in two different Fab structures we have two independent molecules in the asymmetric unit that are in different environments but still have similar structures. In these two examples we see generally the same conformation for all the loops in the two independent copies. However, small changes in the loops may occur because of crystal lattice forces.

Plückthun: You have to choose between two evils. One is to let the system equilibrate, and attempt a co-crystallization. The other is to diffuse in the hapten. In both cases you have problems. In one case you may have to change the crystallization conditions and therefore the contacts between the loops, thereby inducing artificial conformational changes: in the other case you may restrict them. Have you ever done an experiment where you have done both?

Wilson: Yes, we have done both. The crystals often crack when the antigen is diffused in. I feel much happier with co-crystallization than with diffusing in. Also, some ligands are too large to diffuse in, for example in the lysozyme–Fab complexes.

Thornton: In enzymes one gets the impression that part of the active site is nearly always defined by main chain, rigid groups, although part of it may be flexible. In your antibodies are the contacts mainly through side chains or do they involve main chain atoms as well?

Wilson: They use both side chains and the main chain and also use framework residues, for instance in the DB3 structure, Trp-47 of the heavy chain is used (J. H. Arevalo, M. J. Taussig & I. A. Wilson, unpublished).

Hansen: Is there catalytic activity within the crystal?

Wilson: We haven't tested that yet.

Schultz: If you compare the loops and all the different antibody structures that have been solved, are there any generalizations that can be made?

Wilson: There are limited numbers of conformations or canonical structures for L2, L3, H1 and H2; L1 can vary significantly in size and hence in conformation; H3 has no defined set of conformations (see reviews by Chothia & Lesk 1987, Chothia et al 1989).

Burton: CDR3s are extremely variable in length and amino acid sequence, probably much more than people have recognized. In human antibodies we have found CDR3s that vary between six and 22 amino acids.

Benkovic: One antibody we are working with binds *p*-nitrophenol and changes its pK_a. This provided an opportunity to look for conformational changes. We did not see more than simple first-order binding, suggesting that the conformational changes are either silent experimentally or very rapid on our time scale.

Jencks: How fast can you measure?

Benkovic: We can measure conformational changes with rate constants of the order of 800/s.

Hilvert: It might be interesting to look at the physical chemistry of the progesterone-binding antibody where tryptophan moves on binding of the substrate.

Wilson: We have already started binding studies.

Lerner: At the beginning of work on catalytic antibodies, everybody made rules: everyone who solved a crystal structure said all antibodies will be like this one. The immune system is a binding energy machine and anything that can happen will happen, so long as it increases binding energy.

Suckling: What about the relative balance between the hydrophobic-type interactions and hydrogen bonding and ionic interactions? Obviously, the polar groups are going to be particularly important in ionic reactions in which the charge distribution changes during the reaction.

Wilson: Every time there is a charge in the antigen, there are compensating interactions in the antibody. It is not necessarily charged side chains in the antibody that do the neutralizing; backbone amides, carbonyls and side chain hydroxyls may compensate, as in the B13I2 Fab′–peptide structure (Stanfield et al 1990). Otherwise, the hydrophobic interactions play a significant role.

Jencks: Has there been a predominance of relatively non-polar molecules among the Fab molecules subjected to X-ray structure determination?

Wilson: In the solved structures, apart from progesterone and fluorescein, they are predominantly polar ligands such as peptides or proteins.

Lerner: One of the first Fab crystal structures solved was McPC603 with phosphorylcholine bound, where there is strong charge stabilization of the phosphoryl group and of the choline by an aspartate.

Plückthun: There are an arginine and a tyrosine stabilizing the phosphate. The quaternary ammonium ion of choline is charge stabilized by an aspartate, hydrogen-bonded to an asparagine and a glutamate. There is also a lining of hydrophobic residues—leucine, tyrosine and, in some of the phosphorylcholine-binding antibodies, tryptophan. Even this little molecule, phosphorylcholine, uses the two CH_2 groups for hydrophobic binding; the trimethylammonium group is a hydrophobic group too.

Hansen: Do you ever observe a charged amino acid interacting with a peptide bond? I am thinking about the resonance form of peptide bonding using the positive charge on the nitrogen and the negative charge on the oxygen. Is that ever complemented by charged amino acids or do you always have classic hydrogen bonding to a peptide bond?

Wilson: I don't know of any instance where there are those types of charged interactions.

Hansen: It appears that the antibody views a peptide bond more as a neutral structure in terms of the way it interacts with the peptide bond.

Jencks: It acts as a strong dipole.

Hilvert: The oxyanion-binding hole in serine proteases consists of just peptide hydrogen bonds that stabilize the anionic oxygen in the tetrahedral intermediates and transition states.

Hansen: I am not suggesting that it wouldn't work. I was wondering if the peptide bond oxygen ever has a large enough partial negative charge to induce a positively charged amino acid to complement it. The answer so far seems to be that it does not.

Schultz: We have looked at amine oxides as haptens for generating peptidases. We expected an antibody with an oxyanion hole and carboxylate in the combining site. We have had no success to date and have no evidence for complementary electrostatic interactions.

Hilvert: How did you probe for possible interactions?

Schultz: We measured the chemical reactivity of the antibody to a series of substrates, including β-fluoroketone, esters and α-deuteroketones.

Burton: There's a series of crystallized structures of bacterial proteins recognizing carbonate, sulphate and so on. They don't use charged amino acid residues to recognize those anions, they do it all by hydrogen bonding.

Wilson: The same is true for TIM: the phosphate sits on top of the amino end of a 3_{10} helix and is presumably neutralized by the partial dipole.

Thornton: One of the advantages of using those peptide groups rather than charged side chains is that often they are at the ends of helices where they can be rigidly fixed. It is much easier to fix a helix than a side chain that is potentially flexible: the peptides can provide the rigid framework that is needed for the exquisite specificity of those binding proteins.

Jencks: This large dipole on the helix has been discussed very often. It is a huge dipole; I think that's because the helix is a long helix. My understanding is that the net charge at the end of the helix dipole is approximately half a charge, which is exactly the same as in a single peptide or amide. I don't understand why this large dipole is so important or why it is very different from the effect of a single peptide bond.

Thornton: I also have puzzled over why such a small change should be important. I think the reason the amino terminus of a helix is a good binding site is different. At the end of the helix there are the three NH groups together and that's what is special about it. Each NH group can form a slightly stronger hydrogen bond than can an ordinary peptide because the peptide carbonyl itself is involved in a helical hydrogen bond, distorting the electron distribution in the peptide such that the hydrogen bond with the NH group is strengthened. In addition, the position of the helix will be relatively well fixed in space.

Jencks: It is geometric cooperativity at the ends.

Thornton: To move the helix you need a lot more energy than to move a side chain.

Lerner: Many of us want to place cofactors or other things between L1 and L3. If one were to put van der Waals surfaces on a steroid, how many would come within 3–4 Å of a cofactor that was sandwiched with the metal atoms between L1 and L3?

Wilson: Several of the side chains from L1 and L3 would be in contact. For small haptens, the best place to hang a cofactor may be on L2, if it could reach the binding pocket. In solved Fab–hapten structures we have seen so far, it wouldn't then interfere with binding. Most of the interactions we see are between H3 and L3.

Green: There are examples where binding of the antibody can induce a significant conformational change in a hapten or epitope (see Green 1989). Could one estimate the maximum conformational energy change that could be induced by antibody binding? One might devise an experiment where an antibody raised against a twisted molecule would be allowed to bind an analogous planar molecule, then, for example by spectroscopy, determine the degree of twist or bending.

Lerner: Steve, you said we had 20 kcal of binding energy to work with.

Benkovic: If we accept that 10^{-14} is the smallest dissociation constant for antibody–ligand union.

Lerner: The last time I was at a Ciba symposium, Allen Edmundson described the kinds of molecules that like to run into antibody binding sites, and particularly favourable are benzene rings (Edmundson & Ely 1986).

Schwabacher: Another possible reason for flexibility in some of these molecules is that if there were a very rigid binding pocket exactly complementary to the target molecule, there could be slow binding of the product in the pocket. Such a pocket might also allow anything small enough and hydrophobic to bind. If the binding pocket were able to close up somewhat, it would increase selectivity at the expense of losing some affinity for the hapten.

Suckling: How much does the conformation of an antibody change as a function of pH? This is of interest with respect to some of our results on Diels–Alder reactions catalysed by abzymes.

Wilson: I don't know.

Lerner: The classic ways to disrupt antigen–antibody complexes are chaotropic ions, high salt and low pH. In general, low pH is the best. For most antigen–antibody systems, the antigen dissociates at about pH 3. Whether that's ionic or conformational, one doesn't know. The workable range for most antibodies is between pH 5 and 9 and more often pH 6 and 9.

Paul: Could there be a change in the specificity as the pH changes, because of a conformational change?

Lerner: Yes.

Janda: There will be differences in the overall charge of the protein as the pH of the solvent changes. But I assume that between one isomer and another you would see similar binding or no binding at all.

Green: One of the attractions of antibodies is the relatively wide range of pH, temperature and ionic strength over which very strong binding can be observed. It will be interesting to see whether antibody catalysis can also occur over a broad range of reaction conditions.

Fab fragments exhibit the same catalytic activity as the intact IgG. Catalytic activity has also been observed using Fv fragments. Has anyone detected catalysis using heavy or light chains alone?

Plückthun: We have looked at this problem with the McPC603 antibody for which there's no evidence that either the light chain or the heavy chain even binds the transition state analogue. There are structural reasons for that—some of the ionic groups required for binding are on opposite chains: you need the particular arrangement of these two chains for binding. There are also technical problems with heavy chains; many of them have limited solubility.

Paul: We have evidence that purified polyclonal light chains from antibodies raised against vasoactive intestinal peptide are catalytic. Catalysis is probably a property of the monomeric light chain, not of the dimer (Paul et al, this volume).

Janda: When my group initially raised antibody 43C9, which catalysed the hydrolysis of the *p*-nitroanilide and the *p*-nitrophenol ester that Steve Benkovic talked about (this volume), we looked at the Fab fragment. In terms of simple saturation kinetics, for the amide it showed the same kinetic parameters as the intact IgG. K_m and k_{cat} were within a factor of two or less.

Lerner: There's no *a priori* reason to assume that even an F_v fragment wouldn't work.

Schultz: With regard to a different issue, is there any information on bimolecular reactions, particularly on binding constants for antibodies binding to haptens and substrates, as well as reaction rates?

Janda: We have looked at a number of different bimolecular reactions. The first was a phenyl ester with benzylamine: the Michaelis constants were approximately 10^{-3} M for both pieces. Recently, we've looked at the reaction of α-methyl benzyl alcohol with a vinyl ester. We believe the K_m is about $5-10\times10^{-3}$ M for each substrate. The turnover numbers are about 70/minute.

We have looked at a series of haptens where we varied the leaving group, i.e. nitroanilide or α-methyl benzyl phosphonate (Janda et al 1988, 1989). In the latter case, we immunized with racemic material and obtained catalytic antibodies (lipase mimics) for both the R and S isomers; Peter Schultz has also done this with peptides.

Because the so-called lipase class of enzymes can catalyse the reverse reaction of hydrolysis (transesterification) in organic solvents, we have begun to study catalytic antibodies in this type of environment. Initially, we put the lyophilized catalytic antibodies into various organic solvents, then removed the organic solvent and put the antibodies back into buffer. This produced antibodies that had no catalytic activity. We then immobilized some of these antibodies on porous glass beads and found that they retained catalytic activity after being treated with the same sequence of solvents.

We also looked at water miscible solvents such as dimethylformamide, acetonitrile, dioxane and dimethylsulphoxide (DMSO). M.L. Bender used up to 95% dioxane with immobilized α-chymotrypsin and found few differences in its catalytic properties. Above 40% DMSO an antibody will precipitate out, but not if it is immobilized. We found that in 40% DMSO the antibody raised to the α-methyl benzyl phosphonate hapten catalysed the hydrolysis of its homologous ester; the activity was 30–35% of that in pure aqueous solution. This antibody also catalysed the formation of the α-methyl benzyl ester (the reverse reaction) with an impressive k_{cat} of 70/minute. The K_m for both substrates (the vinyl ester and α-methyl benzyl alcohol) was high (about 5–10 mM).

We have done a coupled assay with alcohol dehydrogenase and this catalytic antibody, in which the antibody converts the vinyl ester to acetaldehyde. I don't think anyone had previously done any successful coupled assays of an enzyme with a catalytic antibody. We succeeded because the k_{cat} of this antibody-catalysed reaction is extremely efficient.

Hilvert: In our antibody that catalyses the bimolecular Diels–Alder reaction, maleimide binds with a K_m of about 20 mM. The thiophene dioxide binds with a high K_m: we don't know the actual number because of solubility limitations. For a practical turnover catalyst, you want as high a K_m as possible coupled with as high a k_{cat} as possible, so that you are not limited to working at very low substrate concentrations.

Lerner: Unfortunately, the reaction is so efficient that one cannot get the second order rate constant for the transesterification with the enolic ester.

Janda: In the catalytic antibody field there has always been some comparison with the background. 10^6 rate acceleration over background, especially for the nitroanilide antibody, one thinks is quite good, but it is very poor compared to an enzyme. One rarely compares the rate of an enzyme-catalysed reaction with that of a background reaction, because there is no background rate. For the bimolecular ester formation, there is also no background rate.

Jencks: Is there a concomitant hydrolysis?

Janda: As the vinyl ester gets used, more of the α-methyl benzyl ester is formed. Eventually, the antibody will use this as a substrate and catalyse the reverse reaction.

Schultz: Is there concomitant hydrolysis of the vinyl ester starting material?

Janda: Not really. If we substitute the OH group in the α-methyl benzyl alcohol with an amino group, the compound is no longer a substrate for the antibody. The same is true of hydrazine, but the thiol-substituted compound works, albeit poorly.

Lerner: If one could get the second order rate constant, it would be interesting to see how close it approached the theoretical value. Without a second order rate constant, how could you estimate the effect of molarity?

Page: You could measure the rate constant of the reverse reaction, the hydrolysis, then if you knew the equilibrium constants, you could calculate the second order rate constant.

Schultz: Or you could make use of radioactively labelled compounds. When we were studying phenyl ester hydrolysis in reverse micelles, we couldn't measure a background rate.

References

Benkovic SJ, Adams J, Janda KD, Lerner RA 1991 A catalytic antibody uses a multistep kinetic sequence. In: Catalytic antibodies. Wiley, Chichester (Ciba Found Symp 159) p 4–12

Chothia C, Lesk AM 1987 Canonical structures for the hypervariable regions of immunoglobulins. J Mol Biol 196:901–917

Chothia C, Lesk AM, Tramontano A et al 1989 Conformations of immunoglobulin hypervariable regions. Nature (Lond) 342:877–883

Davies DR, Padlan EA, Sheriff S 1990 Antibody–antigen complexes. Annu Rev Biochem 59:439–473

Edmundson AB, Ely KR 1986 Three-dimensional analyses of the binding of synthetic chemotactic and opioid peptides in the Mcg light chain dimer. In: Synthetic peptides as antigens. Wiley, Chichester (Ciba Found Symp 119) p 107–129

Goodnow CC, Crosbie J, Adelstein S et al 1989 Clonal silencing of self-reactive B lymphocytes in a transgenic mouse model. Cold Spring Harbor Symp Quant Biol 54:907–920

Green BS 1989 Monoclonal antibodies as catalysts and templates for organic chemical reactions. In: Mizrachi A (ed) Advances in biotechnological processes. Alan R Liss, New York 7:359–393

Hartman AB, Mallet CP, Srinivasappa J, Prabhakar BS, Notkins AL, Smith-Gill SJ 1989 Organ reactive autoantibodies from nonimmunized adult Balb/c mice are polyreactive and express non-biased VH gene usage. Mol Immunol 26:359–370

Hilvert D, Gardell SJ, Rutter WJ, Kaiser ET 1986 Evidence against a crucial role for the phenolic hydroxyl of Tyr-248 in peptide and ester hydrolyses catalyzed by carboxypeptidase A: comparative studies of the pH dependencies of the native and Phe-248 mutant forms. J Am Chem Soc 108:5298–5304

Janda KD, Schloeder D, Benkovic SJ, Lerner RA 1988 Induction of an antibody that catalyzes the hydrolysis of an amide bond. Science (Wash DC) 241:1188–1191

Janda KD, Benkovic SJ, Lerner RA 1989 Catalytic antibodies with lipase activity and R or S substrate selectivity. Science (Wash DC) 244:437–440

Logtenberg T 1990 Properties of polyreactive natural antibodies to self and foreign antigens. J Clin Immunol 10:137–140

Mian IS, Bradwell AE, Olson AJ 1991 Structure, function and properties of antibody binding sites. J Mol Biol 217:133–153

Novotny J, Bruccoleri RE, Saul FA 1989 On the attribution of binding energy in antigen–antibody complexes McPC603, D1.3 and HyHEL-5. Biochemistry 28:4735–4749

Padlan EA 1990 On the nature of antibody combining sites: unusual structural features that may confer on these sites an enhanced capacity for binding ligands. Proteins Struct Funct Genet 7:112–124

Paul S, Johnson DR, Massey R 1991 Binding and multiple hydrolytic sites in epitopes recognized by catalytic anti-peptide antibodies. In: Catalytic antibodies. Wiley, Chichester (Ciba Foundation Symposium 159) p 156–173

Phillips DC, Rivers PS, Sternberg MJE, Thornton JM, Wilson IA 1977 An analysis of the three-dimensional structure of chicken triose phosphate isomerase. Biochem Soc Trans 5(B):642–646

Rees DC, Lipscomb WN 1981 Binding of ligands to the active site of carboxypeptidase A. Proc Natl Acad Sci USA 78:5455–5459

Sheriff S, Hendrickson WA, Smith JL 1987 Structure of myohemerythrin in the azidomet state at 1.7/1.3 Å resolution. J Mol Biol 197:273–296

Stanfield RL, Fieser TM, Lerner RA, Wilson IA 1990 Crystal structures of an antibody to a peptide and its complex with peptide antigen at 2.8 Å. Science (Wash DC) 248:712–719

Weber PC, Ohlendorf DH, Wendoloski JJ, Salemme FR 1989 Structural origins of high-affinity biotin binding to streptavidin. Science (Wash DC) 243:85–88

Nuclear magnetic resonance studies of antibody-antigen interactions

Y. Arata

Faculty of Pharmaceutical Sciences, University of Tokyo, Hongo, Tokyo 113, Japan

Abstract. NMR is a useful method for studying antibody–antigen interactions in solution but the large size of the molecules makes interpretation of the data difficult. Multinuclear NMR and a family of switch variant antibodies and their proteolytic fragments are used to circumvent this problem and investigate the process of molecular recognition in antibody function. An Fv fragment has been prepared in high yield from a mouse IgG2a anti-dansyl-L-lysine monoclonal antibody in which the entire C_H1 domain is deleted. ^{13}C NMR resonances are observed for switch variant antibodies selectively labelled by ^{13}C at the carbonyl carbon. Using a double-labelling method, site-specific assignments have been completed for all the methionine resonances in these antibodies. Comparison of the NMR spectra for intact antibodies and those of their proteolytic fragments suggests that carbonyl carbon chemical shift data can provide information on the way in which information is transmitted through different domains on antigen binding. Selective labelling of the antibody with ^1H, ^{13}C or ^{15}N followed by nuclear Overhauser effect spectroscopy has identified some of the residues involved in antigen binding and indicated that the structure of the Fv fragment is significantly affected by antigen binding.

1991 Catalytic antibodies. Wiley, Chichester (Ciba Foundation Symposium 159) p 40–54

X-ray diffraction is potentially the most direct method of studying antigen–antibody interactions at the atomic level. Crystal structures of a variety of antigen–antibody complexes have been obtained using Fab and, more recently, Fv fragments of antibodies (Davies et al 1990). Problems inherent in the interpretation of X-ray data obtained from proteins include intermolecular interactions in the crystal and the presence of large amounts of inorganic salts. Therefore care should be taken when estimating conformational changes associated with antigen binding on the basis of X-ray diffraction data.

Nuclear magnetic resonance (NMR) is another method that can be used to investigate antigen–antibody interactions at the atomic level of resolution. An important feature of NMR is that it can give information about protein structures *in solution*; thus it can be more useful than X-ray diffraction for measuring possible changes in the conformation of an antibody in the presence and absence of antigen. We have shown that ^{1}H NMR provides useful information on the flexible portion of immunoglobulin G (IgG) molecules and their proteolytic fragments (Arata et al 1980, Endo & Arata 1985, Ito & Arata 1985). However, the large size of these molecules makes it virtually impossible to follow the standard procedure used for the structural analysis of smaller proteins (Wüthrich 1986).

A major problem in applying NMR to the study of proteins such as antibodies or antibody fragments is the observation and assignment of individual NMR resonances, because of the large number of amino acid residues that give different chemical shifts in the NMR spectrum. Appropriate methods need to be developed to extract as much relevant information as possible.

Anglister & Zilber (1990) have reported a ^{1}H NMR study of antigen–antibody interactions using the Fab fragment of an anti-peptide monoclonal antibody. In their analyses both the antigen and the Fab fragment were selectively deuterated, and two-dimensional transferred nuclear Overhauser effect (NOE) difference spectroscopy was used. On the basis of the difference data, they showed how the antigen interacts with some of the amino acid residues in the antigen-binding site. However, information on possible changes in the conformation of the site on antigen binding was not obtained, because they measured spectral changes after adding more antigen to an Fab fragment already saturated with the antigen. This was necessary to circumvent the spectral complexity caused by the large size of the Fab fragment.

In this paper we shall describe an approach using NMR which attempts to overcome problems due to the size of the proteins and to understand the processes of molecular recognition in a variety of antibody functions. Two important features of this approach are (1) the use of a family of switch variant antibodies and their proteolytic fragments, including Fv, and (2) the use of stable isotope-aided multinuclear NMR spectroscopy.

Switch variant anti-dansyl mouse IgG antibodies

Switch variant antibodies have identical V_H, V_L and C_L chains joined to different C_H chains (Fig. 1). Dangl et al (1982) generated a family of switch variant mouse monoclonal antibodies against dansyl linked to keyhole limpet haemocyanin; these were selected and cloned using a fluorescence-activated cell sorter. This family of antibodies contains an interesting IgG2a variant in which the heavy chain is 10 kDa smaller than normal (Dangl et al 1982, Oi et al 1984). We have shown (Igarashi et al 1990) that the entire C_H1 domain is deleted in this variant, hereafter designated IgG2a(s).

We are taking advantage of this family of switch variant antibodies that possess identical light chains and an identical V_H domain in two ways. First, we use the family of homologous antibodies for spectral assignments. Then those

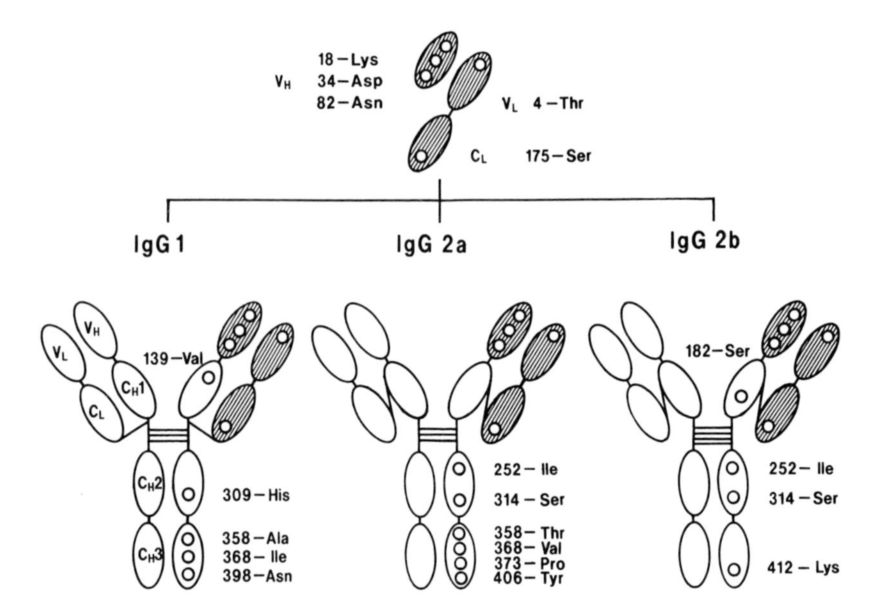

FIG. 1. Schematic diagram of the distribution of all the Met residues in the switch variant anti-dansyl IgG1, IgG2a and IgG2b mouse antibodies. The Met residues are represented by their position numbers in the amino acid sequences; the amino acid residue directly linked by a peptide bond to each Met residue is shown (e.g. 18-Lys indicates the dipeptide sequence Met-18–Lys-19). The V_H, V_L, and C_L domains, which have identical sequences in all the proteins shown, are hatched and are drawn separately at the top.

results are used as a basis for consideration of domain–domain interactions, especially the way in which information is transmitted through different domains on antigen binding. ^{13}C NMR resonances are observed for switch variant antibodies selectively labelled by ^{13}C at the carbonyl carbon. Site-specific resonance assignments are thus possible for all the carbonyl carbons of Met residues in these switch variant antibodies.

The IgG2a(s) antibody can be cleaved by clostripain to give the corresponding Fv fragment in high yield (Takahashi et al 1991). Fv is a heterodimer of one V_H and one V_L domain; it is the smallest antigen-binding unit (M_r 25 000). Two-dimensional NOE spectroscopy is being used to analyse the molecular mechanism of antigen recognition by anti-dansyl Fv fragments.

Stable isotope-aided NMR spectroscopy of antibodies and their proteolytic fragments

Selective deuteration was first used by Jardetzky and co-workers to obtain information by ^1H NMR about the structure of proteins in solution at the

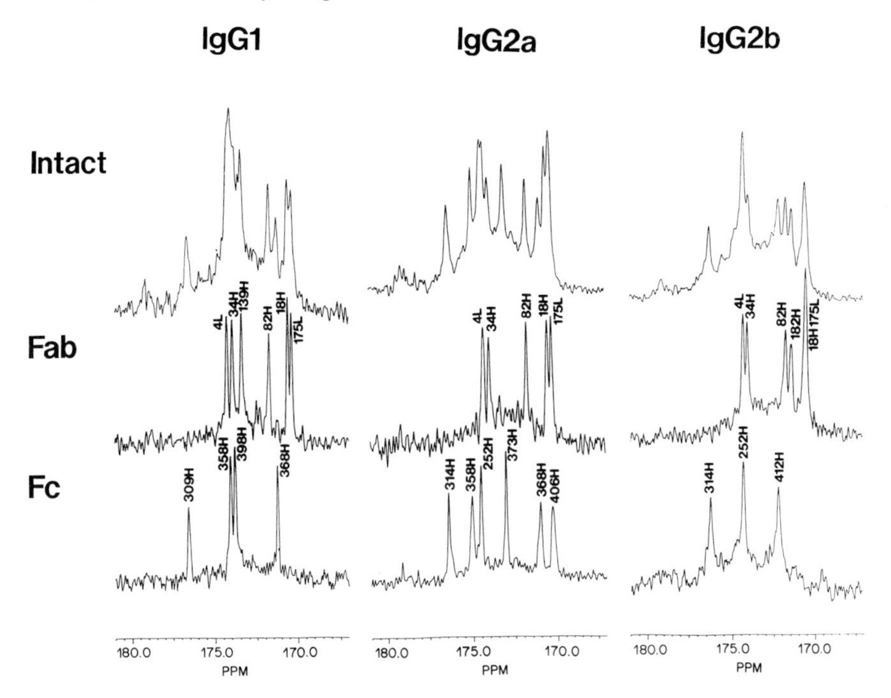

FIG. 2. The 100 MHz ^{13}C NMR spectra of switch variant anti-dansyl IgG1, IgG2a and IgG2b antibodies and their proteolytic fragments Fab and Fc. The protein concentrations and the pH of the sample solutions were in the ranges 0.1–0.4 mM and 7.1–7.5, respectively. The probe temperature was 30 °C. 32 000–180 000 transients were accumulated for the measurements. For each measurement, 32 000 data points were used with a spectral width of 24 000 Hz and a delay of 0.3 s.

atomic level (Markley et al 1968). Multinuclear NMR techniques, which use proteins singly or doubly labelled with ^{2}H, ^{13}C and ^{15}N, are now applied extensively to the determination of detailed solution structures of proteins of M_r up to 20 000 (Markley 1989).

Using antibodies labelled with [1-^{13}C]Met at the carbonyl carbon we are able to observe all the carbonyl carbon resonances separately and to make site-specific assignments to all these resonances (Kato et al 1989a, 1991). The digestion products Fv, Fab* (V_H, V_L and C_L), Fab and Fc were used for fragment-specific and domain-specific resonance assignments. We have used a method described by Anglister et al (1985) to assign resonances to specific chains within the antibody. The double-labelling method of Kainosho & Tsuji (1982) was used to make the site-specific spectral assignments.

We shall now summarize the process of spectral assignment for the IgG2b antibody.

Spectral assignment of the IgG2b antibody

Figs. 1 and 2 show that there is a straightforward correspondence between the observed ^{13}C resonances of the intact antibody and those of its proteolytic fragments Fab and Fc. The observed resonances were assigned to the heavy or light chain by the chain recombination experiment. The light chain resonances originating from Met-4L and Met-175L can be identified[†]. The site-specific assignment of the Met-4L resonance by the double-labelling method led automatically to the assignment of the Met-175L resonance. The Met-18H, Met-34H and Met-82H resonances were assigned by combinations of different types of doubly labelled proteins. The resonance for Met-182H, which is the only Met residue in the C_H1 domain, was assigned by comparing spectra from the Fv and Fab fragments of IgG2b. The Fc fragment contains three Met residues at positions 252H, 314H and 412H. The Met-252H resonance was assigned by the double-labelling method (Fig. 3); the site-specific assignment of the Met-412H resonance was also done this way (Kato et al 1989a). From these results the assignment of the Met-314H resonance could be deduced. By combining the results of these experiments, site-specific assignments of all the carbonyl carbon ^{13}C resonances have been completed for the IgG2b antibody.

Similar experiments have enabled us to assign all the carbonyl carbon ^{13}C resonances for the switch variant antibodies IgG1, IgG2a and IgG2(s).

We are using ^{13}C labelling of Met, Cys, His, Trp and Tyr residues of the family of anti-dansyl switch variant antibodies to investigate the process of molecular recognition in antibody function.

NMR spectra of intact antibodies and proteolytic fragments

In most cases the ^{13}C spectra observed for the intact IgG antibodies are simply a superposition of those of their proteolytic fragments. This presumably reflects the independent nature of the functional domains of the antibody with respect to methionyl carbonyl carbon resonances. However, our ^{13}C study has shown that even in the switch variant IgG proteins, which have identical V_L, V_H and C_L domains, significant chemical shifts are observed for some of the ^{13}C resonances originating from the V_H and C_L domains. A significant example is the Met-175L resonance observed for the Fab* and Fab fragments. This is shifted upfield by 1.6 p.p.m. by deletion of the C_H1 domain. The Met-175L (C_L) resonance also gives different chemical shifts for Fab fragments with homologous C_H1 domains. We suggest that the carbonyl carbon chemical shift data may be used to determine the way in which information is transmitted through different domains on antigen binding.

[†]Met-4 and Met-18 in the light and heavy chains are designated Met-4L and Met-18H, respectively. Similar notations are used for all Met residues. For the numbering system used in this paper see Kabat et al (1987).

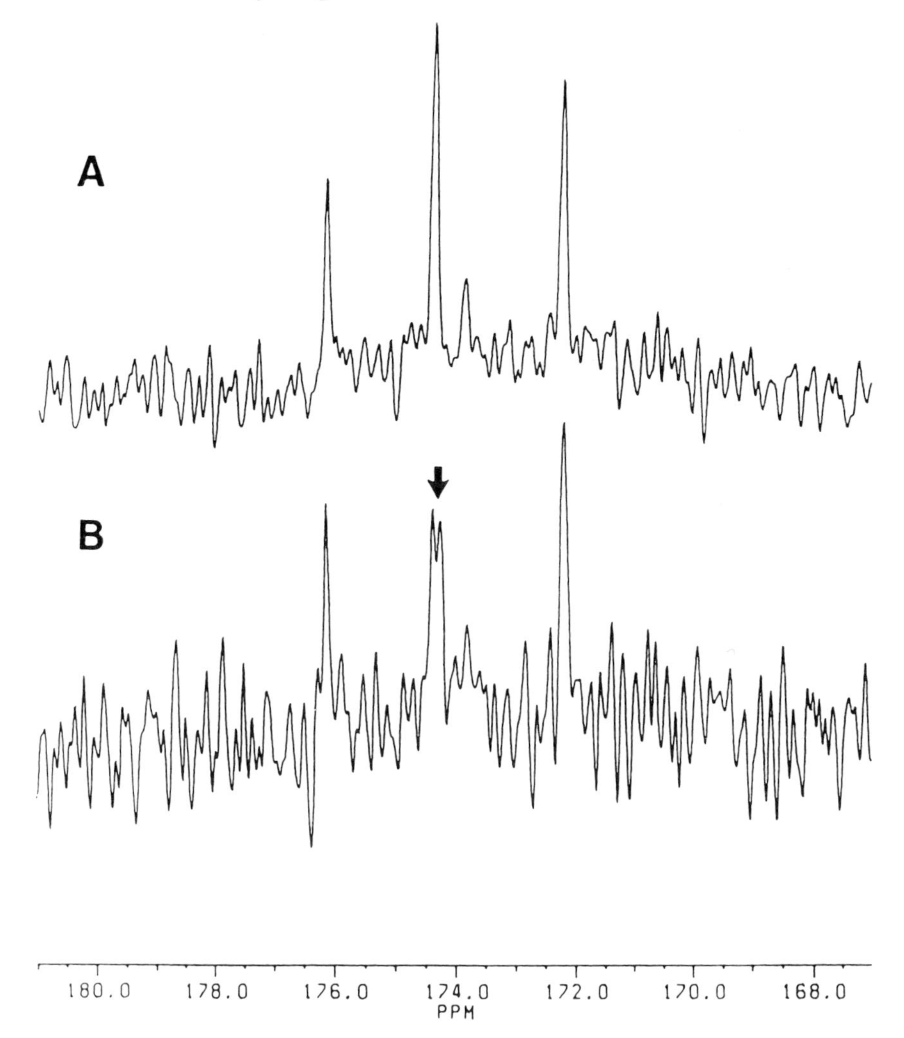

FIG. 3. The 100 MHz ^{13}C NMR spectra of the Fc fragment of the IgG2b antibody. (A) Fc singly labelled with [1-^{13}C]Met. (B) Fc doubly labelled with [1-^{13}C]Met and [^{15}N]Ile. The probe temperature was 45 °C. Other experimental conditions are as in Fig. 2. Resonance line split into a doublet because of spin coupling with ^{15}N is indicated by the arrow.

Antigen–antibody interactions in the Fv fragment

Inbar et al (1972) were the first to prepare Fv fragment by pepsin digestion of a mouse myeloma IgA(λ_2) protein, MOPC 315. Attempts to obtain Fv fragments by proteolysis of whole IgG molecules with other types of light chain have not yet succeeded (Sharon & Givol 1976). The expression and secretion

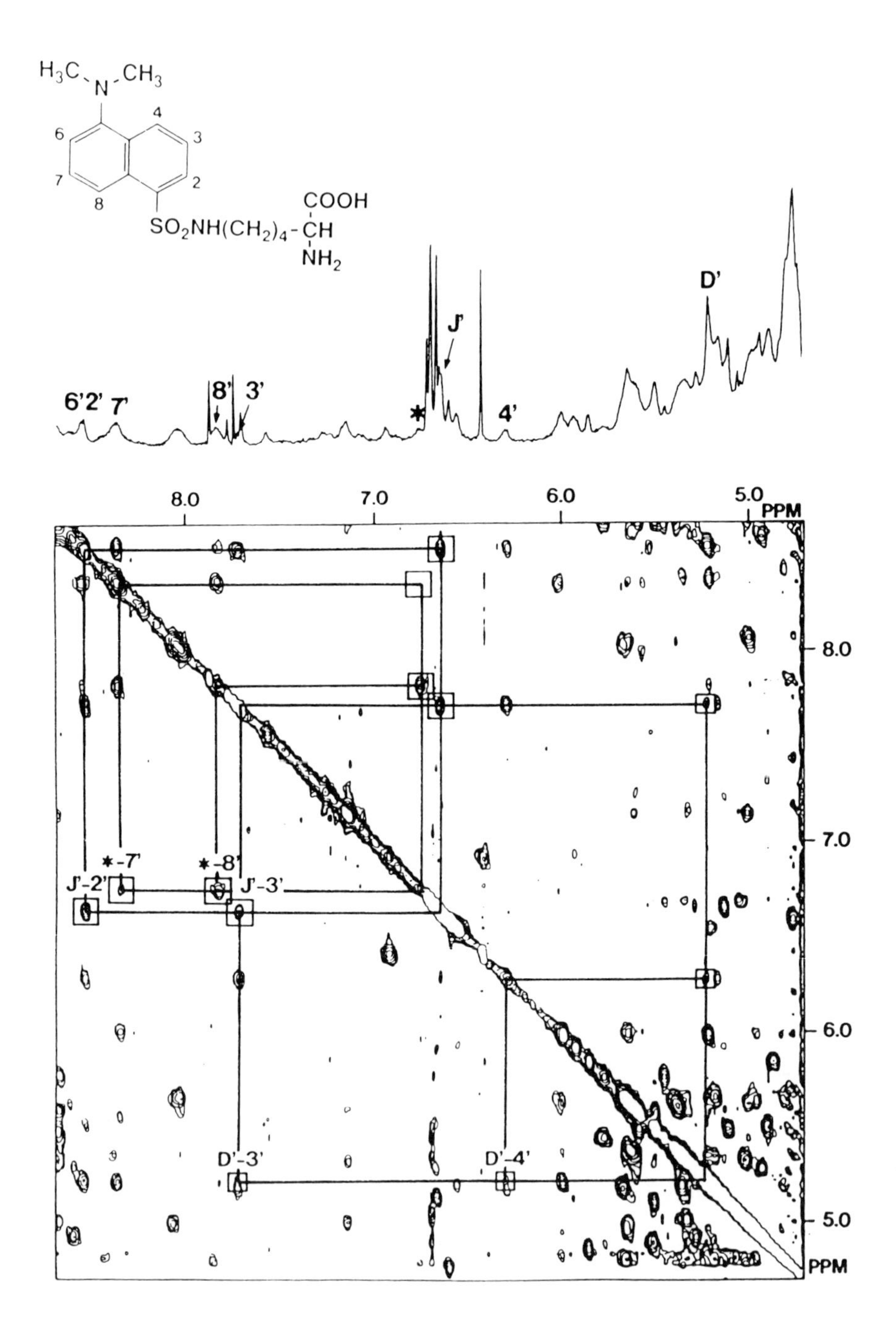
H_3C N CH_3
4
6
3
7
2
8
COOH
$SO_2NH(CH_2)_4$-CH
NH_2
D'
J'
6'2' 7'
8' 3'
*
4'
8.0
7.0
6.0
5.0
PPM
*-7'
*-8'
J'-2'
J'-3'
D'-3'
D'-4'
8.0
7.0
6.0
5.0
PPM

of Fv fragments by *Escherichia coli* have recently been reported (Skerra & Plückthun 1988, Ward et al 1989). X-ray structural analyses of Fv fragments derived by gene technology have been started (Glockshuber et al 1990, Boulot et al 1990). Preliminary NMR studies of an Fv fragment expressed in myeloma cells (Riechmann et al 1988) have been described by Wright et al (1990). They outlined a general strategy for spectral assignments by the combination of stable isotope labelling and a variety of two-dimensional techniques.

As mentioned above, treatment of the short chain IgG2a(s) antibody with clostripain gives the corresponding Fv fragment in high yield. We have established a procedure for selectively labelling the antibody with ^{2}H, ^{13}C or ^{15}N (Kato et al 1989a,b, 1991). Thus, a variety of Fv analogues are now available in large quantities for structural analysis by NMR.

NOE spectra of an Fv analogue in which all aromatic protons except for His C2′-H and Tyr C3′, 5′-H had been deuterated were measured in the presence of different amounts of the antigen, dansyl-L-lysine. From these spectra it was possible to assign all the ring proton resonances for the dansyl group bound to the Fv fragment. We also obtained information about His and Tyr residues of the Fv fragment in the presence and absence of the antigen. At least two Tyr residues and one of the amide groups are directly involved in binding the antigen.

The anti-dansyl Fv fragment used in this study contains 13 Tyr residues, which we have designated A–M. Tyr D and Tyr J of the Fv–dansyl complex give NOE cross-peaks with protons 3 and 4 and protons 2 and 3 of the dansyl ring, respectively (Fig. 4). In addition, NOE spectroscopy shows an interaction of protons 7 and 8 of the dansyl ring with an amide proton of the Fv fragment. This proton is resistant to proton–deuterium exchange on prolonged incubation of the Fv fragment in D_2O, suggesting that the amide group is shielded from the solvent and is involved in the network of hydrogen bonding.

The chemical shifts of the resonances originating from Tyr D and Tyr J are significantly affected by addition of the antigen (data not shown). The Tyr D resonance is shifted upfield by 1.38 p.p.m; proton 4 of the dansyl ring exhibits an upfield shift of 1.15 p.p.m. As NOE shows that proton 4 of the dansyl ring is close to the Tyr D ring (Fig. 4), the large chemical shifts observed on complex formation are probably induced by the mutual ring current of the aromatic rings of Tyr D and dansyl. From these observations, it is quite likely that in the Fv–dansyl complex the aromatic rings of Tyr D and dansyl are stacked

FIG. 4. (*Opposite*) The 500 MHz one-dimensional (*upper*) and two-dimensional (*lower*) NOE spectra of [^{2}H]Fv fragments of IgG measured in the presence of antigen (dansyl-L-lysine; *upper left*). The [^{2}H]Fv:dansyl-L-lysine molar ratio was 1:1. A mixing time of 150 ms was used for the measurements. Virtually identical results are obtained with a mixing time of 80 ms (data not shown). NOE cross-peaks observed between Tyr and dansyl-L-lysine in the Fv-dansyl-L-lysine complex are shown by squares and are labelled D′-3′, D′-4′, J′-2′, J′-3′, *-7′ and *-8′.

adjacently. On the basis of these results we suggest that Tyr D, Tyr J and the amide group form a hydrophobic environment which interacts with the antigen.

The antigen-combining site of the antibody molecule is constructed by three complementarity-determining region (CDR) loops from the heavy chain and three from the light chain. In the anti-dansyl Fv fragment used in this study, there are seven Tyr residues in the CDR loops: Tyr-58 and Tyr-59 in CDR 2 of the heavy chain; Tyr-96, Tyr-97, Tyr-99 and Tyr-104 in CDR 3 of the heavy chain, and Tyr-32 in CDR 1 of the light chain. ^{13}C NMR spectra were measured using Fv analogues in which the carbonyl carbon of a Tyr or Trp residue was specifically labelled with ^{13}C. The chemical shifts for the carbonyl carbon resonances of Tyr-97, Tyr-99, Tyr-104, Trp-101 and Trp-105 are selectively affected to a large extent by the addition of the antigen (A. Odaka et al, manuscript in preparation). Thus for this anti-dansyl Fv fragment, CDR 3 of the heavy chain is primarily responsible for the antigen binding. The recombination experiment (see above) showed that resonances D and J originate from the heavy chain; we therefore suggest that they come from the CDR 3 loop of the heavy chain. In order to assess a contribution to the observed chemical shift from other amino acid residues in the antigen-combining site, we are doing similar NOE spectroscopy experiments using different types of deuterated Fv analogues.

Three NOE cross-peaks involving His and Tyr residues are observed in the absence of the antigen: two of these peaks disappear on antigen binding. The linewidth for Tyr D is broad in the absence of the antigen and becomes much sharper in its presence. On the basis of these NOE spectroscopy data, we suggest that the structure of the Fv fragment is significantly affected by the antigen binding.

Conclusion

The variety of NMR methods described in this paper should be most effective in the structural analyses of antibodies when they are used in combination with X-ray crystallography and computer modelling. The availability of genetically engineered antibody fragments (Fab and Fv) will facilitate these studies.

Acknowledgements

We thank Professor L. A. Herzenberg, Stanford University, and Dr V. T. Oi, Becton Dickinson Immunocytometry Systems, for generously providing us with the switch variant hybridoma cell lines and also for making the amino acid sequence data of the variable regions available to us prior to publication. I am most grateful to my colleagues and collaborators, particularly I. Shimada, K. Kato, H. Takahashi, T. Igarashi, C. Matsunaga, H. Kim, A. Odaka, S. Yamato and W. Takaha for their contributions to this work. This research was supported by Special

Coordination Funds for Promoting Science and Technology from the Science and Technology Agency, Japan.

References

Anglister J, Zilber B 1990 Antibodies against a peptide of cholera toxin differing in cross-reactivity with the toxin differ in their specific interactions with the peptide as observed by ^{1}H NMR spectroscopy. Biochemistry 29:921–928

Anglister J, Frey T, McConnell HM 1985 NMR technique for assessing contributions of heavy and light chains to an antibody combining site. Nature (Lond) 315:65–67

Arata Y, Honzawa M, Shimizu A 1980 Proton nuclear magnetic resonance studies of human immunoglobulins: conformation of the hinge region of the IgG1 immunoglobulin. Biochemistry 19:2784–2790

Boulot G, Eisele J-L, Bentley GA et al 1990 Crystallization and preliminary X-ray diffraction study of the bacterially expressed Fv from the monoclonal anti-lysozyme antibody D1.3 and of its complex with the antigen lysozyme. J Mol Biol 213:617–619

Dangl JL, Parks DR, Oi VT, Herzenberg LA 1982 Rapid isolation of cloned isotype switch variants using fluorescence activated cell sorting. Cytometry 2:37–43

Davies RD, Padlan EA, Sheriff S 1990 Antibody–antigen complexes. Annu Rev Biochem 59:439–473

Endo S, Arata Y 1985 Proton nuclear magnetic resonance study of human immunoglobulins G1 and their proteolytic fragments: structure of the hinge region and effects of a hinge-region deletion on internal flexibility. Biochemistry 24:1561–1568

Glockshuber R, Steipe B, Huber R, Plückthun A 1990 Crystallization and preliminary X-ray studies of the V_L domain of the antibody McPC603 produced in *Escherichia coli*. J Mol Biol 213:613–615

Igarashi T, Sato M, Katsube Y et al 1990 Structure of a mouse immunoglobulin G that lacks the entire C_H1 domain: protein sequencing and small-angle X-ray scattering studies. Biochemistry 29:5727–5733

Inbar D, Hochman J, Givol D 1972 Localization of antibody-combining sites within the variable portions of heavy and light chains. Proc Natl Acad Sci USA 69:2659–2662

Ito W, Arata Y 1985 Proton nuclear magnetic resonance study on the dynamics of the conformation of the hinge segment of human G1 immunoglobulin. Biochemistry 24:6467–6474

Kabat EA, Wu TT, Reid-Miller M, Perry HM, Gottesman KS 1987 Sequences of proteins of immunological interest, 4th edn. US Department of Health and Human Services, National Institutes of Health, Washington, DC

Kainosho M, Tsuji T 1982 Assignment of the three methionyl carbonyl carbon resonances in *Streptomyces subtilisin* inhibitor by a carbon-13 and nitrogen-15 double-labeling technique. A new strategy for structural studies of proteins in solution. Biochemistry 24:6273–6279

Kato K, Matsunaga C, Nishimura Y, Waelchli M, Kainosho M, Arata Y 1989a Application of ^{13}C nuclear magnetic resonance spectroscopy to molecular structural analyses of antibody molecules. J Biochem (Tokyo) 105:867–869

Kato K, Nishimura Y, Waelchli M, Arata Y 1989b Proton nuclear magnetic resonance study of a selectively deuterated mouse monoclonal antibody: use of two-dimensional homonuclear Hartmann-Hahn spectroscopy. J Biochem (Tokyo) 106:361–364

Kato K, Matsunaga C, Igarashi T et al 1991 Complete assignment of the methionyl carbonyl carbon resonances in switch variant anti-dansyl antibodies labeled with [1-^{13}C]methionine. Biochemistry 30:270–278

Markley JL 1989 Two-dimensional nuclear magnetic resonance spectroscopy of proteins: an overview. Methods Enzymol 176:12–64
Markley JL, Putter I, Jardeztky O 1968 High-resolution nuclear magnetic resonance spectra of selectively deuterated staphylococcal nuclease. Science (Wash DC) 161:1249–1251
Oi VT, Hsu C, Hardy R, Herzenberg LA 1984 Hybridoma antibody-producing switch variants: a variant lacking the CH1 domain. In: Beers RF Jr, Bassett EG (eds) Cell fusion: gene transfer and transformation. Raven Press, New York
Riechmann L, Foote J, Winter G 1988 Expression of an antibody Fv fragment in myeloma cells. J Mol Biol 203:825–828
Sharon J, Givol D 1976 Preparation of Fv fragment from the mouse myeloma XRPC-25 immunoglobulin possessing anti-dinitrophenyl activity. Biochemistry 15:1591–1594
Skerra A, Plückthun A 1988 Assembly of a functional immunoglobulin Fv fragment in *Escherichia coli*. Science (Wash DC) 240:1038–1041
Takahashi H, Igarashi T, Shimada I, Arata Y 1991 Preparation of the Fv fragment from a short-chain mouse IgG2a anti-dansyl monoclonal antibody and use of selectively deuterated Fv analogues for two-dimensional ^{1}H NMR analyses of the antigen–antibody interactions. Biochemistry 30:2840–2847
Ward ES, Güssow D, Griffiths AD, Jones PT, Winter G 1989 Binding activities of a repertoire of single immunoglobulin variable domains secreted from *Escherichia coli*. Nature (Lond) 341:544–546
Wright PE, Dyson HJ, Lerner RA, Riechmann L, Tsang P 1990 Antigen–antibody interactions: an NMR approach. Biochem Pharmacol 40:83–88
Wüthrich K 1986 NMR of proteins and nucleic acids. Wiley, New York

DISCUSSION

Lerner: May we take it from these very beautiful results that one could examine all antigens using these kinds of labelling technologies?

Arata: By following the procedure described here (Kato et al 1991, Takahashi et al 1991), it should be possible to determine what part of an antigen binds to what part of an antibody.

Lerner: If you had an antigen that was not very rigid and you wanted simply to see to what extent the conformation approximates that of the transition state, what would be the time frame for collecting the data for such an experiment?

Arata: That is a very interesting but difficult question. If you start with virtually no NMR structural data, you may have to spend a couple of years on such experiments.

Lerner: How dependent would the experiment be on the off-rate of the antigen? To study the antigen, assuming that diffusion will control the on-rate, what kind of binding constants are needed?

Arata: That depends on the information that one wants. For transfer NOE, which Jacob Anglister (1990) has been using, you need an off-rate of typically 500 sec^{-1}. In other cases, you require a very stable antigen–antibody complex to obtain information about the way in which the antigen and antibody interact.

Lerner: The transfer NOE method is compatible with average off-rates, however?

Arata: Yes.

Schultz: Have you carried out any spin label mapping experiments to find out which Met residues in the Fab fragment are in close proximity to the hapten?

Arata: A couple of groups, including Raymond Dwek's group at Oxford, tried it, but that was before the new isotope-labelling technique was introduced to this field. I should like to do this type of spin label experiment using labelled proteins, which would give us information about epitope mapping. I would also like to try paramagnetic lanthanide probes for the same purpose.

Burton: Professor Arata, are your NMR experiments really telling you how the antigen is held? You say that a tyrosine is perturbed in CDR 3 when antigen is added to antibody. What you really know is that the tyrosine is in a new environment when the antigen binds. How much can you truly deduce from chemical shift data—the fact that the tyrosine has moved or that it's in the proximity of the antigen?

Arata: Using NMR, one of the most difficult things is to interpret the chemical shift data in a quantitative way. The chemical shifts of the carbonyl carbon resonances can be influenced by a number of factors, including the electrostatic interactions between the carbonyl group and the surroundings. In 3–4 years, when we have accumulated sufficient data, we shall be able to say more about the interpretation of the chemical shift. I would like to emphasize that the qualitative data we are now getting give us invaluable information about what is happening in the antigen binding site.

Burton: You are not doing two-dimensional NMR and getting the three-dimensional structure of the antigen in the antibody combining site; you are using essentially a difference technique. You are saying that something has happened, but you can't say what.

Arata: That is why we have been trying to combine ^{2}H, ^{13}C and ^{15}N NMR. In the case of ^{13}C, it's at present difficult to interpret our results quantitatively, but even in that case we can say for sure that the CDR 3 region is primarily responsible for antigen binding.

Burton: Surely you can't say from that NMR spectrum that most of the binding energy is in the CDR 3?

Arata: We can, because we are not observing just one resonance; we are observing more than 50 resonances, from different parts of the entire Fv molecule. Can you imagine a situation in which a different part of the molecule is responsible for the antigen binding when only the CDR 3 resonances are affected to a great extent? That is unlikely! Moreover, we have also used proton NMR, and that is what you call a two-dimensional approach (Takahashi et al 1991).

With an antibody it is not necessary to do the structural determination of the whole protein, because you know the structure of the framework from X-ray crystallography combined with suggestions from computer modelling, as Dr Thornton will discuss in her paper (Thornton, this volume). We have obtained information about CDR regions by multinuclear NMR techniques. By combining results from these techniques, we shall get a reasonable picture of the antigen-binding site.

Plückthun: Let us consider the amount of work that is necessary to obtain a completely assigned spectrum of the whole Fv region, which is the prerequisite for any three-dimensional structure determination. If you take (1) the number of experiments that are necessary when you label each of the 20 amino acids in turn, and then do a couple of double-labelling experiments to get assignments, and (2) the random ^{13}C or ^{15}N labelling of all the amino acids using the bacterial expression system, how would you say these alternatives compare for getting the structure?

Arata: If we get a reasonable amount of material expressed in *E. coli* and biosynthetically labelled with ^{13}C and ^{15}N, that approach will eventually lead to the right conclusion. At present it is difficult to deal with large molecules (25 kDa) by multinuclear NMR, but we should be able to do so in the future. Antibody molecules are so diverse that establishing a way to get information about specific parts of the molecule, for example CDR 3, is essential.

Plückthun: In principle, is it possible, using your method of labelling one amino acid after another, eventually to assemble enough information to assign the complete proton NMR spectrum?

Arata: In principle, yes. But, that is not the way we should go. As I said, structural determination of the whole protein is not necessary for an antibody because the structure of the framework is known from X-ray crystallography. We should concentrate on the antigen-binding site, and this is what NMR is for. I should mention that it is difficult to incorporate, for example, ^{15}N-labelled aspartic acid and glutamic acid (Kato et al 1991).

Leatherbarrow: NMR-derived structures are intrinsically different from 'crystal' structures. There are regions of NMR structures that are well defined and regions that are not. Typically, regions of regular secondary structure are well defined by NMR and regions that do not have defined secondary structure, like loops (and this applies directly to the antibody combining site) are not well defined by NMR, even in small proteins. So even if you can solve a complete solution structure of an antibody molecule by NMR, the framework region may be well defined while the antibody combining site is not.

Arata: The important thing is that we can clearly observe NMR resonances when the part of the molecule we are interested in is flexible. This part would not presumably give any definite electron density by X-ray crystallography. Of course, we will not be able to get any definite NOE spectrum for this part of the molecule. We would, however, be able to describe qualitatively how the antigen-binding site changes its dynamic structure in the absence and presence of antigen. This kind of information cannot be obtained from X-ray crystallography.

Leatherbarrow: But getting information about things that don't have a defined secondary structure is what NMR does worst. Antibodies are not only large molecules, but the regions that are of most interest are those that NMR will define least well.

Arata: I would like to repeat the statement I have made. In discussing the dynamic nature of the antigen-binding site in solution, it would certainly not be appropriate to rely on X-ray crystallographic data. We should use X-ray and NMR, which are complementary.

Burton: Are the hypervariable loops really flexible, or not? Ian Wilson said (this volume) that the crystallographic evidence was that they were not flexible, because in two different crystal forms the loop was in essentially the same conformation.

Wilson: Yes. There are quite a few hydrogen bonds within the loops as well, so they are not as flexible as was originally thought.

Thornton: Basically, the hypervariable loops are not flexible. The existence of 'canonical' structures of the loop suggests a fixed main chain framework, but does not prove this. Some of the side chains are flexible, but most are fixed in one position by hydrogen bonding and specific interactions.

Lerner: People are confusing what is flexible relative to another protein with what is flexible within the protein. Ian Wilson and Janet Thornton have said that if you look at C_H3 regions, they are in very different positions in different molecules, but within an antibody, the loops are not so flexible.

Burton: Ian Wilson said that the CDR 3 region did move when antigen bound. That is within the same protein. If there is movement of the order of 1 Å, hydrogen bonds will be disrupted and so on.

Wilson: Not necessarily, the loops can move in a segmental-type motion, as I described (this volume), without breaking hydrogen bonds within the loop.

Burton: I would like to know whether the hypervariable loops are flexible and then get frozen when the antigen binds, or whether, as Ian Wilson suggests, that's not the case and the antigen goes into the combining site and causes the whole loop to move, which must involve sacrificing a fair amount of energy.

Thornton: The movement of loops when antigen binds to the combining site isn't very great. The evidence suggests rather minor changes involving principally side chains. Data on protein–antibody complexes show little movement of the loops. In general, the loops are well defined, but this will depend on the antibody in question: some of the very long loops may be rather flexible.

Wilson: But even L1 was thought to be a lot more flexible than it turns out to be. Three of our structures have long L1 loops and these are in similar conformations.

Burton: These questions of loop flexibility and movement on antigen binding are very interesting. They have a lot to do with antibody design, and how difficult this will be.

Paul: Has the particular antibody that you have used, Professor Arata, been looked at by X-ray crystallography?

Arata: We have just started working with Yoshinori Satow's group in our department using the anti-dansyl Fv; we shall see results within a couple of years.

References

Anglister J 1990 Use of deuterium labelling in NMR studies of antibody combining site structure. Q Rev Biophys 23:175–203

Kato K, Matsunaga C, Igarashi T et al 1991 Complete assignment of the methionyl carbonyl carbon resonances in switch variant anti-dansyl antibodies labelled with [1-^{13}C]methionine. Biochemistry 30:270–278

Takahashi H, Igarashi T, Shimada I, Arata Y 1991 Preparation of the Fv fragment from a short-chain mouse IgG2a anti-dansyl monoclonal antibody and use of selectively deuterated Fv analogues for two-dimensional ^{1}H NMR analyses of the antigen–antibody interactions. Biochemistry 30:2840–2847

Thornton JM 1991 Modelling antibody combining sites: a review. In: Catalytic antibodies. Wiley, Chichester (Ciba Found Symp 159) p 55–71

Wilson IA, Stanfield RL, Rini JM et al 1991 Structural aspects of antibodies and antibody–antigen complexes. In: Catalytic antibodies. Wiley, Chichester (Ciba Found Symp 159) p 13–39

Modelling antibody combining sites: a review

Janet M. Thornton

Biomolecular Structure and Modelling Unit, Biochemistry and Molecular Biology Department, University College, London WC1E 6BT, UK

Abstract. The combining site in an antibody is built up from six hypervariable loop regions, three from the light chain and three from the heavy chain. Three-dimensional structures have been elucidated by X-ray crystallographic studies for a variety of combining sites and for four protein–antibody complexes. Since sequence determination is relatively straightforward, whilst structure determination is often difficult and time consuming, it would be useful to be able to predict the structure of a combining site from its sequence. Knowledge of the structure would facilitate modifications of antibodies for specific aims. The structure of an antibody combining site for which only the sequence is known can be modelled on the basis of homology with a protein of known structure. The most demanding steps are modelling of the hypervariable loops and inclusion of the amino acid side chains. Final energy refinement of the model is achieved by conventional energy minimization techniques, but simulated annealing promises to be a powerful way of predicting side chain conformation. Four different groups have modelled antibody combining sites and compared their predictions with observed structures: for the main chain conformations resolutions of less than 1 Å have been achieved. Future developments will increase the efficiency of the modelling procedures and permit accurate predictions of side chain conformations.

1991 Catalytic antibodies. Wiley, Chichester (Ciba Foundation Symposium 159) p 55–71

Antibodies have the unique capacity amongst proteins to recognize almost any foreign molecule. This is achieved by extensive sequence and structural variation in the antibody combining site. Crystallographic three-dimensional structures for more than 20 antibodies are available (Table 1). However, in the sequence database there are more than 1720 immunoglobulin sequences for which there are no three-dimensional data. The structures show that the eight-stranded β-barrel core of the immunoglobulins is well conserved (Chothia & Lesk 1987). The variation in structure (as in sequence) is limited to the combining site. Therefore, over the last 10 years there have been intense efforts to model the structure of a new sequence on the basis of known structures. With the increasing

TABLE 1 Brookhaven files: July 1990

Immunoglobulin		*Author*	*Date deposited*	*Resolution (Å)*
1MCG	Ig B-J intact MCG	A. Schiffer, A. Edmundson et al	1978	2.3
1REI	Ig B-J fragment (V-dimer) REI	O. Epp, R. Huber	1976	2.0
2RHE	Ig B-J fragment (V-monomer) RHE	W. Furey, B. C. Wang, C. S. Yoo, M. Sax	1983	1.6
2FBJ	IgA Fab (κ) J539	T. Bhat, E. Padlan, D. R. Davies	1989	2.6
1MCP	IgA Fab (κ) McPC603	Y. Satow, G. Cohen, E. Padlan, D. R. Davies	1984	2.7
2MCP	IgA Fab (κ) McPC603/phosphocholine	E. Padlan, G. Cohen, D. R. Davies	1984	3.1
2FB4	IgG_1 Fab (λ) KOL	M. Marquart, R. Huber	1989	1.9
3FAB	Ig Fab (′) new	R. Poljak	1981	2.0
4FAB	Ig 4-4-20 Fab/fluorescein	A. Edmundson et al	1989	2.7
1F19	Fab F19.9 (mouse)	R. Poljak et al	1988	2.8
1FC1	Ig Fc (human)	J. Deisenhofer	1981	2.9
1FC2	Ig Fc-fragment B complex	J. Deisenhofer	1981	2.8
IPFC	IgG PFC fragment	L. M. Amzel	1981	3.1
2IG2	IgG_1 (λ) KOL	M. Marquart, R. Huber	1989	3.0
Complex (Ig-antigen)				
2HFL	HYHEL-5 Fab/lysozyme complex	S. Sheriff, D. R. Davies	1987	2.5
3HFM	HYHEL-10 Fab/lysozyme complex	E. Padlan, D. R. Davies	1988	3.0

use of monoclonal antibodies in medicine as well as in all aspects of biochemistry, knowledge of antibody structure offers several advantages. For example, it becomes possible to design rational modifications to improve an antibody's affinity for a given target or its catalytic efficiency. Ultimately, the dream is to design *ab initio* novel antibodies with a given specificity or enzymic activity.

In this paper I shall describe the methods available for modelling combining sites, the accuracies achieved and possibilities for the future.

Structure of combining sites

Antibody structures available in the Brookhaven Data Base (Bernstein et al 1977) are listed in Table 1 with their resolution. These include antibody–hapten and antibody–protein complexes. Recently, the structure of an antibody–peptide complex has also been solved (Stanfield et al 1990).

The combining site of the antibody is made up from six loop regions, three from the light chain (called L1, L2 and L3) and three from the heavy chain (H1, H2 and H3) (Fig. 1, Table 2). These loops join β-strands, which form the immunoglobulin β-barrel core. In protein structure nomenclature four loops (L2, L3, H2, H3) are part of β-hairpins (i.e. the adjacent strands are hydrogen bonded together) whilst two loops (L1, H1) form β-arches connecting strands in the two opposing sheets (Fig. 1). The length of the loops is somewhat variable but does seem to be held within limits. For example, in $V_{\varkappa}$ chains, over 95% of L2 loops are the same length, whilst the length of L1 varies by up to seven residues (Chothia et al 1989). Various methods are used to define which residues form part of the framework conserved region and which form the loops. Originally, sequence alone was used to define the hypervariable regions; however, as more structures become available the core is now often defined, using least squares fitting algorithms, as that part of the structure which is superposable within specified limits. Recent developments of more flexible algorithms to find structural equivalence between residues permit an improved definition of the framework regions (Taylor & Orengo 1989, Sali & Blundell 1990). In practice, different groups use different definitions for modelling, which makes comparisons difficult! The other problem in assessing the accuracy of the models is the accuracy of the crystallographic coordinates. In proteins the electron density for loop regions (especially the long ones) is often weak and difficult to interpret, but it is these very regions that are being modelled and for which accurate data are required (see below).

Conventionally, loop regions in proteins have been referred to as 'random coil' structures with no apparent regularity. We have shown that short loops are tightly constrained by end-loop positions and orientations, and adopt a limited set of preferred loop conformations (Thornton et al 1988). For example, in β-hairpins, four 'structural families' have been observed (Sibanda & Thornton 1985) occurring throughout the database, which suggests structural stability and

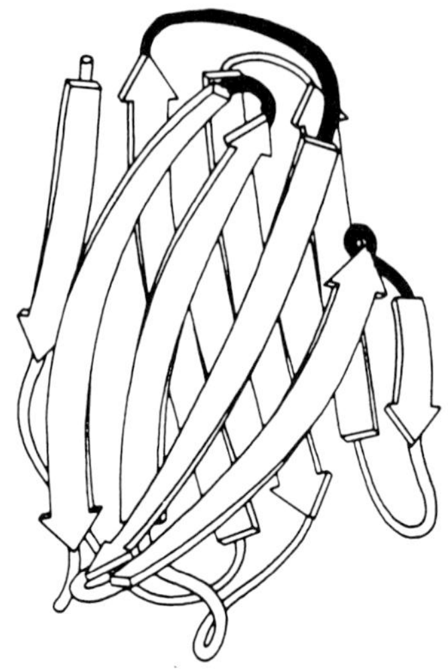

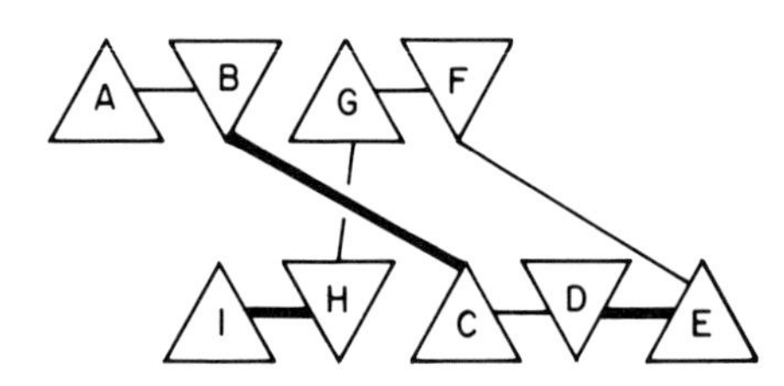

FIG. 1. Topology of the V_L domain of an antibody. The three hypervariable loops which contribute to the combining sites are indicated by heavy black lines. The broad arrows represent β-strands. The topology of the two β-sheets is represented below the ribbon diagram using triangles to indicate each strand. The triangles point towards the C terminus of each strand.

convergent evolution. Common structural changes resulting from loop sequence modifications are summarized in Fig. 2.

In 1977 Kabat et al noticed that certain residues within the loops were conserved and suggested that they may be structurally important. Padlan & Davies (1975) had already found that gross changes in sequence in the loops sometimes had little effect on conformation. Later studies on more structures (de la Paz et al 1986, Chothia & Lesk 1987) strengthened this observation and led to the suggestion that at least five of the six hypervariable regions adopt a limited number of canonical conformations. Chothia & Lesk (1987) identified critical residues which through packing, hydrogen bonding or conformational

TABLE 2 Canonical structures of hypervariable regions

Loop	*Number of canonical structures*	*Number of residues*	*% Known human sequences that fit*
L1	4	6	—
		7	60
		13	5
		12	5
L2	1	3	95
L3	3	6	90
		6	—
		5	2
H1	1	7	50
H2	4	3	15
		4	1
		4	40
		6	15

From Chothia et al (1989).

flexibility are thought to be primarily responsible for the main chain conformation of the loops. Fig. 3, adapted from Chothia et al (1989), illustrates the limited set of conformations suggested for the L1, L2, L3 and H1, H2 loops. In contrast, the H3 loop is quite long and adopts a greater variety of structures. These observations have direct relevance for modelling the combining site.

The overall shape of the combining site, which determines specificity and affinity, is determined not only by the main chain fold of the loops but also by the disposition of the side chains. Although attention has recently centred on accurately predicting the main chain fold as the first stage, a useful model should include side chains.

Methods of modelling antibody structure

Antibody modelling is part of the wider field of homology modelling (see Blundell et al 1987 for a review) in which one protein is modelled on the basis of an homologous structure. There are several distinct steps in such modelling:

(i) Define the structurally conserved regions or core;
(ii) Align the sequences;
(iii) Construct the optimal core;
(iv) Model the hypervariable loops onto the framework;
(v) Model the amino acid side chains onto the main chain conformation;
(vi) Final energy refinement.

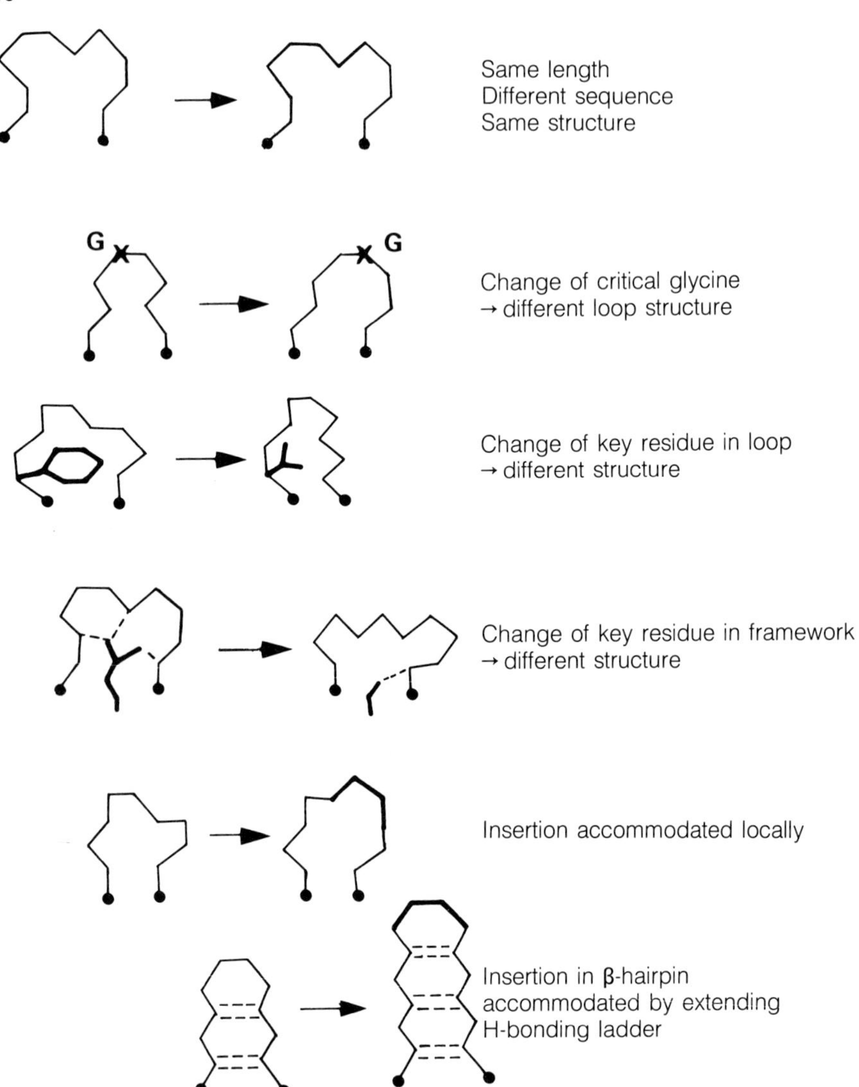

FIG. 2. Common structural changes resulting from loop sequence modifications.

For the antibody combining sites, steps (iv)–(vi) are most demanding and will be described further.

Loop modelling

When the optimal β-barrel framework has been constructed, a variety of methods can be used to model the loops.

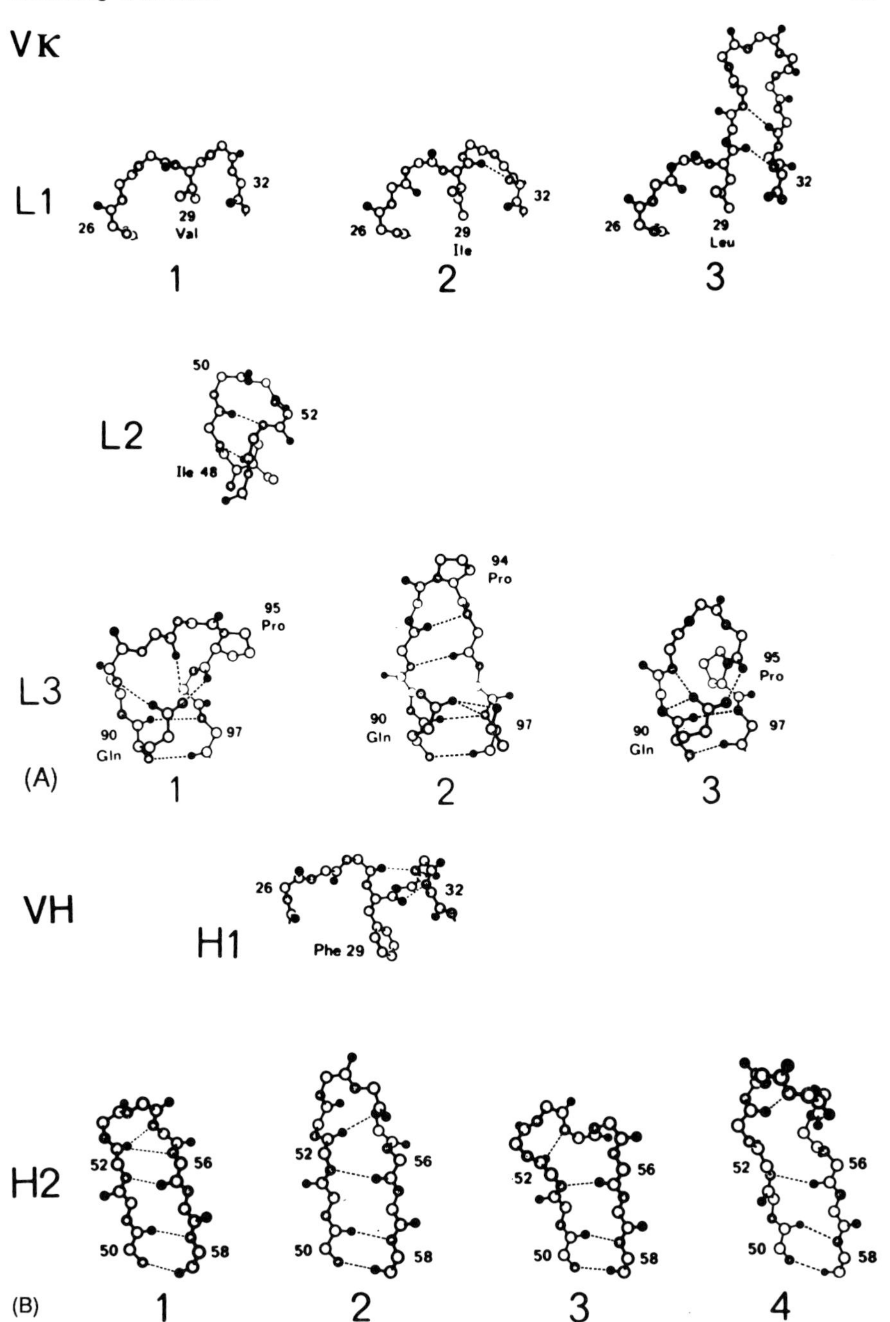

FIG. 3. Canonical structures of the hypervariable loops. A) V_κ light chains. B) Heavy chains. From Chothia et al (1989).

Loop extraction from a database of known structures

The end-point coordinates of the loop are used to extract all loops of the correct length from a database of known structures. Given the importance of 'critical' residues it is often useful to impose sequence constraints on the loop searches. Glycines, buried hydrophobic or hydrophilic residues which participate in critical hydrogen bonds should be 'conserved' to maintain the loop structure. The method of Jones & Thirup (1986) provides an elegant, fast search procedure. The loop is then fitted onto the framework using various geometric algorithms.

Global searching in ϕ,ψ space

All conformations of the loop in ϕ,ψ space are generated and each is tested to see if it is sterically allowed and will close the loop (eg Moult & James 1986). 'Best' loops are then chosen, usually according to an energy-based criterion. This approach is only really feasible for shorter loops. The method of Go & Scheraga (1970) may also be implemented to close the loop.

Energy minimization/simulated annealing/molecular dynamics

In principle it is possible to construct a loop structure in a single conformation and then use molecular mechanisms/dynamics algorithms to explore all states of that loop and select the one with the lowest energy (Bruccoleri et al 1988, Fine et al 1986). There are two problems. Firstly, the old methods of energy minimization did not explore conformational space adequately. The new simulated annealing techniques look much more promising. Secondly, current energy parameters do not distinguish between conformers sufficiently, so that the observed structure is not usually the one with lowest energy (Novotny et al 1988).

Side chain modelling

In homologous structures the conformation of the side chains is usually conserved (Summers et al 1987). Therefore the general rule for determining the χ value for the side chains is to model the new amino acid using χ_1, χ_2 . . . etc as appropriate from the old residue. If the old residue is a glycine, or shorter than its replacement, no χ values are available and the new residue is usually modelled either fully extended or preferably in the most common conformation for that residue type in a loop (McGregor et al 1987). Global searching of side chain conformations has also been used (Moult & James 1986). In the antibody combining site, many of the side chains will be accessible to solvent and may be flexible or adopt alternative conformations. This will complicate realistic modelling, especially when a complex is formed and side chain mobility may be important.

Energy refinement

All modelled structures, regardless of the methodology used to obtain them, require energy refinement, not only to regularize the geometry of the model but also to relieve bad contacts and optimize interaction energies. Conventional energy minimization moves the model by only approximately 0.5 Å root mean square deviation from its starting point. Simulated annealing is much more powerful and can explore a wider range of conformational space. It has been used with great success to predict the conformation of the side chains for flavodoxin given the backbone geometry (C. Lee, personal communication).

Results obtained using different modelling techniques

How good are the models of the antibody combining sites? A survey will be given of results presented in the literature using these various methods. Three aspects will be considered:

(1) Main chain conformation of the loops;
(2) Relative disposition of the loops;
(3) Modelling of side chains.

Chothia, Lesk and colleagues have used the knowledge-based approach, restricted strictly to immunoglobulin structures as the 'source of knowledge'. They modelled the structures of four immunoglobulin combining sites, whose structures were subsequently solved (Chothia et al 1989). Their method is non-automatic: each loop region is modelled independently *in isolation*. They examine the new sequence to determine whether it is the same size as any hypervariable region of known structure, and whether its sequence contains the set of residues responsible for a known conformation. Side chain conformation is considered only briefly. They found that 19 of the 24 (4×6) loops can be modelled using the 'canonical' library (one L3 loop and all the H3 loops could not be modelled with this method). Of the 19 predicted loops the root mean square deviation for all main chain (N, C_α, C, O) atoms ranged from 0.3–1.3 Å (average = 0.76 Å), which represents excellent agreement (see Fig. 4A). The positioning of loops relative to the framework regions was less good, with C_α positions deviating by 0.5–3.0 Å (see Fig. 4B). This reflects both changes in the framework region relative to the reference structure and shifts in the loops.

Bruccoleri et al (1988) deleted all the hypervariable loops from two immunoglobulins and reconstructed them using the conformational search program CONGEN (Bruccoleri & Karplus 1987). A protocol was developed whereby the loops were generated in a specific sequential order (L2-H1-L3-H2-H3-L1), chosen according to the positions of the loops relative to the framework. The shorter loops (L2-H1-L3) do not interact with each other and provide a natural basis for the construction of the remaining longer loops (H2-H3-L1) which

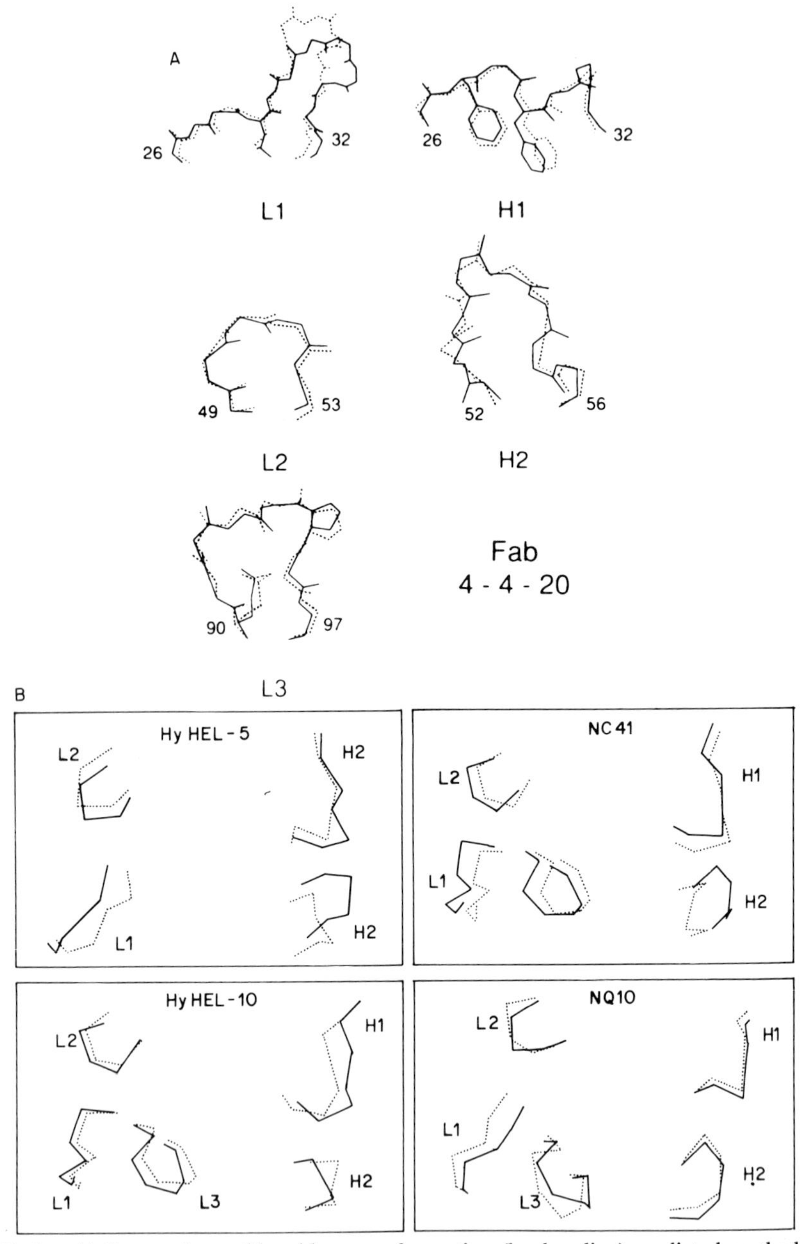

FIG. 4. A) Comparison of local loop conformation (broken line) predicted on the basis of canonical loop structures with observed structure (continuous line). Average main chain deviations were 0.3–1.3 Å. B) Relative positions of the hypervariable regions in predicted and observed structures. The fit is not so good; C_α positions deviated by up to 3 Å. From Chothia et al (1989).

extend further from the framework. The conformational search algorithm generates a set of loop conformations which satisfy the fixed end-point conditions. Potential energies are then used to distinguish between 'good' and 'bad' conformations, which largely depend on short-range loop–loop interactions. The 'best' loop is selected to be incorporated into the final model on the basis of energy and low solvent accessibility. The longer loops (H2 and L1) had to be constructed in stages. This procedure is very computer-intensive, taking 20 days computer user time on a MicroVax II to model the combining site of McPC603. It gives root mean square deviations from the crystal structure of 1.4–1.7 Å for the polypeptide backbone and 2.4–2.6 Å for all atoms.

Fine et al (1986) modelled four of the loops (H1, H3, L1, L3) in antibody McPC603. They generated random loop conformations followed by either energy minimization or dynamics plus minimization! Interestingly, they proposed that the two shortest loops (H1 and L2) could be modelled using only alanines, suggesting that these loops fold independently of sequence variation (cf. only one canonical structure found for these loops by Chothia et al 1989). They found that L3 showed significant conformational variability and no single structure could be chosen. The fourth loop, H3, required side chain interactions to fix the conformation into that observed in the crystal structure (cf. Chothia et al could not identify canonical structures for this loop).

In contrast, Martin et al (1989) used a fully automated approach which combined knowledge-based loop extraction, global ϕ,ψ loop generation and energy evaluation. They attempted to model each combining site loop *in situ*. They used the entire Brookhaven Protein Databank as a source of loop structures and selected loops on the basis of distance constraints derived from antibody structures. Having selected the best loop and corrected its sequence, they 'deleted and reconstructed' the distal portion of the loop, or that portion surrounding any insertion or deletion, using the conformational search program CONGEN. The side chains were positioned using conformational searching. The conformations generated were screened using a modified potential energy function and solvent accessibility criteria. To test their method, Martin et al (1989) modelled two antibodies whose structures were later solved: HyHel-5 in the presence of the antigen lysozyme, and Gloop2. All 12 loops were modelled, half on the basis of antibody loop structures, three from non-antibody loops and three solely by conformational search techniques. Only three loops were not reconstructed in part using CONGEN. Comparison of the predicted and observed structures showed root mean square superpositions for backbone (N, C_α, C) atoms from 0.9–1.5 Å (average = 1.0 Å) and for all atoms (including side chains) from 1.0–2.9 Å (average = 2.0 Å). Predictions for the Gloop2 loops are shown in Fig. 5. They also modelled loop L3 in HyHel-5 in the absence of lysozyme and predicted that its preferred conformation changes: crystallographic data are needed to assess the validity of this. Unfortunately, no attempt was made to assess systematically the effect of reconstruction using CONGEN.

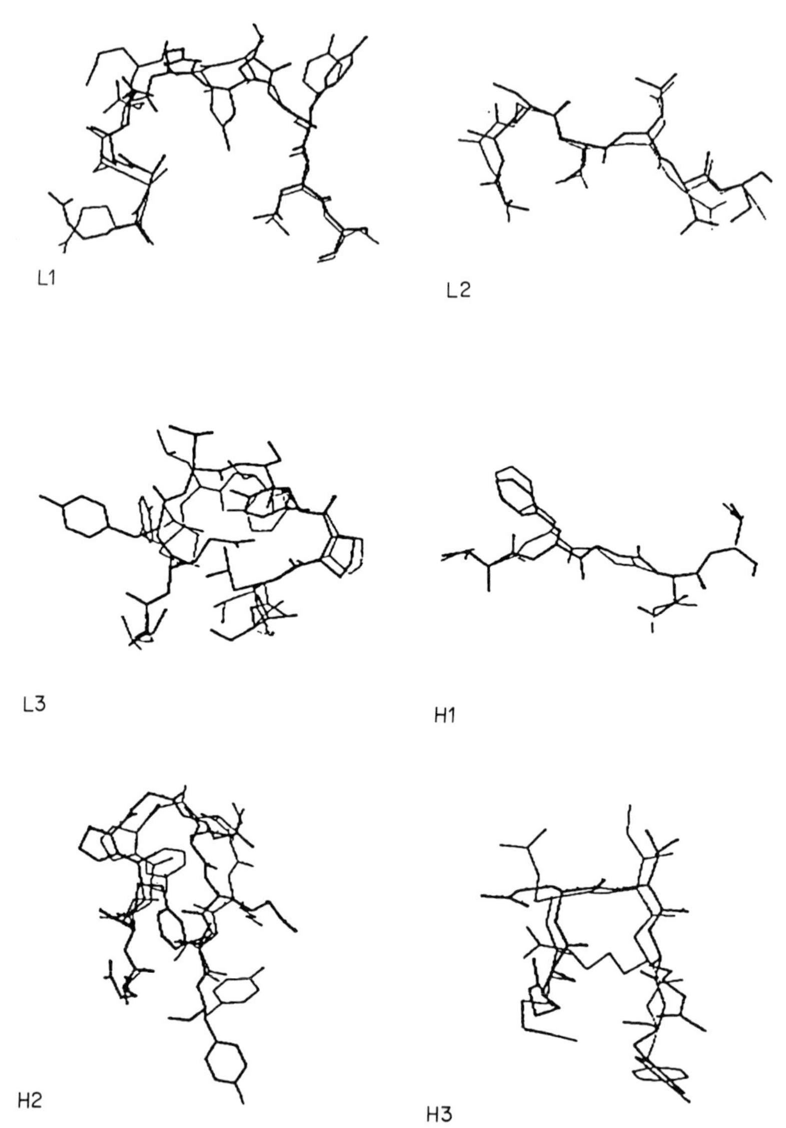

FIG. 5. Conformation of the hypervariable loops in antibody Gloop2 predicted by Martin et al (1989) (thin lines) and the observed conformation (thick lines).

Discussion

A direct comparison of the results from these different methods is not valid since they address different modelling problems; they use loops of different length, and they compare their agreement to the observed structures using different atoms. However, the summary of results, presented in Table 3, does

TABLE 3 Comparison of modelling methods

	Chothia et al 1989 Four structures (main chain)	*Martin et al 1989 Two structures (backbone)*	*Bruccoleri et al 1988 Two structures (backbone)*
L1	0.75 (6–13)	1.1 (10)	1.6 (5–12)
L2	0.75 (3)	0.7 (7)	1.2 (6)
L3	0.5 (5)	1.35 (9)	0.95 (5–6)
H1	1.0 (7)	0.95 (5)	0.9 (5)
H2	0.9 (3–6)	1.25 (10)	1.85 (7–9)
H3	—	1.2 (4–7)	1.05 (4–8)
All loops	0.3–1.3	0.9–1.5	0.6–2.6
All atoms	—	1.0–2.9	1.4–4.1

The values are the root mean square deviations from the crystal structures, in Ångströms. The numbers of residues in the loops modelled are given in parentheses.

indicate that it is now possible to model the main chain conformations of antibody combining sites to better than 1 Å resolution. This compares favourably with low resolution crystal structure determination. The table shows that the simplistic canonical structure method works well for predicting the loop conformation, although the more *ab initio* approach of Bruccoleri et al also does reasonably well. In general, the shorter loops are easier to model, especially if there is only one canonical structure. A few comments may be made with reference to future developments:

(1) A modelling package needs to be fully automated, requiring as little human intervention as possible.
(2) The canonical structure hypothesis has been shown to be valid for a limited number of structures; further examples would strengthen the hypothesis. However, it is a very powerful predictor which implicitly includes the influence of the conserved framework, thus equivalent antibody loops should be used where possible as a basis for modelling. Critical residues should be identified computationally so that the appropriate canonical structure can be identified.
(3) Improved methods to attach loops onto the framework and allow the whole structure to 'relax' need to be developed. Simulated annealing is one such powerful tool, which will hopefully improve the models.
(4) Molecular recognition of the antigen involves the side chains of the combining site. So far there are few studies on the prediction of the conformation and mobility of such surface residues and crystallographic data on conformational changes that occur when antigen binds are much needed. Conventional energy calculations do not adequately predict the observed structure. Future developments of a more sophisticated 'hydrophobic force' are required.

Attempts to predict antibody–antigen complex formation are still in their infancy and the rational design of antibodies with a predefined specificity remains a goal for the future. However, the first step towards this goal, the successful prediction of antibody combining sites, has been made. The experimental techniques for constructing, expressing and extracting an antibody of any sequence are now available. We must develop the theoretical approaches to complement these experimental techniques which will allow intelligent design of new antibodies.

References

Bernstein FC, Koetzle TF, Williams GJB et al 1977 The protein databank: a computer based archival file for macromolecular structure. J Mol Biol 112:535–542

Blundell TL, Sibanda BL, Sternberg MJE, Thornton JM 1987 Knowledge-based prediction of protein structures and the design of novel molecules. Nature (Lond) 326:347–352

Bruccoleri RE, Karplus M 1987 Prediction of folding of short polypeptide segments by uniform conformational sampling. Biopolymers 26:137–168

Bruccoleri RE, Haber E, Novotny J 1988 Structure of antibody hypervariable loops reproduced by a conformational search algorithm. Nature (Lond) 335:564–568

Chothia C, Lesk AM 1987 Canonical structures for the hypervariable regions of immunoglobulins. J Mol Biol 196:901–917

Chothia C, Lesk AM, Tramontano A et al 1989 Conformation of immunoglobulin hypervariable regions. Nature (Lond) 342:877–883

de la Paz P, Sutton BJ, Darsley MJ, Rees AR 1986 Modelling of the combining sites of anti-lysozyme monoclonal antibodies and of the complex between one of the antibodies and its epitope. EMBO (Eur Mol Biol Organ) J 5:415–425

Fine RM, Wang H, Shenkin PS, Yarmush DL, Levinthal C 1986 Predicting antibody hypervariable loop conformations. II: Minimisation and molecular dynamics studies of McPC603 from randomly generated loop conformations. Proteins Struct Funct Genet 1:342–362

Go N, Scheraga HA 1970 Ring closure and local conformational deformation of chain molecules. Macromolecules 3:178–187

Jones TA, Thirup S 1986 Using known substructures in protein model-building and crystallography. EMBO (Eur Mol Biol Organ) J 5:819–822

Kabat EA, Wu TT, Bilofsky H 1977 Unusual distributions of amino acids in complementarity-determining (hypervariable) segments of heavy and light chains of immunglobulins and their possible roles in specificity of antibody-combining sites. J Biol Chem 252:6609–6616

Martin ACR, Cheetham JC, Rees AR 1989 Modelling antibody hypervariable loops: a combined algorithm. Proc Natl Acad Sci USA 86:9268–9272

McGregor M, Islam SI, Sternberg MJE 1987 Analysis of the relationship between side-chain conformation and secondary structure in globular proteins. J Mol Biol 198:295–310

Moult J, James MNG 1986 An algorithm for determining the conformation of polypeptide segments in proteins by systematic search. Proteins Struct Funct Genet 1:146–163

Novotny J, Rashin AA, Bruccoleri RE 1988 Criteria that discriminate between native proteins and incorrectly folded models. Proteins Struct Funct Genet 4:19–30

Padlan EA, Davis DR 1975 Variability of three-dimensional structure in immunoglobulins. Proc Natl Acad Sci USA 72:819–823

Sali A, Blundell TL 1990 Definition of general topological equivalence in protein structures. J Mol Biol 212:403–428
Sibanda BL, Thornton JM 1985 β-Hairpin families in globular proteins. Nature (Lond) 316:170–174
Stanfield RL, Fieser TM, Lerner RA, Wilson IA 1990 Crystal structures of an antibody to a peptide and its complex with peptide antigen at 2.8 Å. Science (Wash DC) 248:712–719
Summers NL, Carlson WD, Karplus M 1987 Analysis of side-chain orientations in homologous proteins. J Mol Biol 196:175–198
Taylor WR, Orengo C 1989 Protein structure alignment. J Mol Biol 208:1–22
Thornton JM, Sibanda BL, Edwards MS, Barlow DJ 1988 Analysis, design and modification of loop regions in proteins. BioEssays 8:63–69

DISCUSSION

Benkovic: From energy minimization calculations one sometimes gets an idea how many conformations are nearly equal in energy. What do those calculations tell one about conformational flexibility? Are we looking at antibody molecules that have several states?

Thornton: In terms of complementarity-determining regions (CDRs), some do and some do not. Those with only one canonical structure appear to have only one possible state and that is presumably the one with the minimum energy.

Benkovic: For those antibodies with three or four conformational states, how many kilocalories of free energy would separate one from another?

Thornton: I suspect less than five. The absolute values derived from energy calculations are not very accurate.

Benkovic: Does one have any idea how rapid the interconversion events are, from these calculations? Do you get any idea from NMR by looking at time-dependent nuclear Overhauser effect spectroscopy?

Thornton: To my knowledge, that has not been done.

Wilson: One problem with the modelling is the change in pairing between V_H and V_L, which can vary significantly, as I described for Allen Edmundson's BV04-01 Fab (Wilson et al, this volume).

Thornton: I agree; it would alter completely the orientation of the CDRs relative to each other, which would change radically the structure of the combining site. The inability to model the core or framework accurately remains a serious limitation. Alternative methods using distance geometry are being developed.

Lerner: Is modelling useful, given that you can't look at CDR 3 in the heavy chain?

Thornton: The only approach that does not attempt to model CDR 3 is that used by C. Chothia, A. Lesk and colleagues, based on canonical structures. The global search and energy-based techniques can predict structures for CDR 3, although it is difficult to identify the correct fold.

Modelling can provide guidance for experimental work. For example, in the design of mutagenesis experiments it can suggest which residues are close to each other and may interact. It also depends how much the antigen of interest interacts with this particular H3 loop in the antibody.

Burton: If you want to take one or two of the loops for a particular function, this method can give information on that. Fine et al (1986) found that they could replace the L2 loop essentially with polyalanine—that is amazing.

Thornton: Polyalanine in L2 would be expected to have the canonical backbone structure, since the prediction depends on the end-point constraints and polypeptide chain geometry rather than specific side chains. However, the specificity of the antibody will depend on these side chains.

Burton: Is any of this useful for the study of catalysis? In catalysis 1 Å is a huge distance.

Thornton: I agree; for catalysis, predictions need to be accurate to less than 1 Å.

Lerner: It's not as though catalysis were independent of binding. If you could solve the binding parameters, you would learn a lot about catalysis.

Hilvert: Ian, in the antibody structures that you have solved, do all the loops fit the canonical picture?

Wilson: From studies we did with A. Lesk and C. Chothia, L2, L3, H1 and H2 fit quite well; L1 is wrongly predicted but that prediction was based on only one other Fab with a long L1 loop, which has a poorly defined L1 (Chothia et al 1989). All our L1s agree with each other. H3 has no canonical structure.

Plückthun: Only a small fraction of the known lengths of L1 have actually been observed in the structures determined so far. From the Kabat sequence database on mouse κ light chains, only class I, class V and class VI have ever been studied crystallographically. We have made a mutant belonging to class II and analysed it crystallographically (B. Steipe, R. Huber & A. Plückthun, unpublished work) but, as far as I know, class III and class IV have never been crystallized. They are different in length and I think they would have to give rise to a different canonical structure, because two things of different lengths cannot have the same canonical structure.

Thornton: Then you have to go back to basics. We obviously need more structures to assess how valid these canonical structures are. I always thought of the CDRs as being very long and variable in their conformation and they are surprisingly not.

Wilson: The point about the coordinates is well taken: it seems that as people refine their structures, these become more like the canonical structures.

Thornton: We have been looking at the reliability of protein structures. It is amazing that as the structures are defined at higher resolution, everything gets more regular.

Lerner: But when you refine the structure don't you put an energy minimization function into your algorithm?

Thornton: It depends on the method. Most of the structures in the database are calculated using the Hendrikson–Konnert least squares 'PROLSQ' refinement program, which does not include strong energy-based constraints. Other algorithms incorporate strong energy constraints, which are reflected in the resulting coordinates.

Above 2 Å resolution, there are insufficient data to define the backbone ϕ, ψ angles accurately without added energy constraints. Below 2 Å there are enough data to define the backbone electron density clearly, and the ϕ, ψ angles are found to be regular.

Wilson: In some cases, the predicted structures may be more accurate than some of those that have been built, particularly with lower resolution structures. For example, the refined structure of D1.3 anti-lysozyme Fab is now closer to the predicted structure (Chothia et al 1989).

References

Chothia C, Lesk AM, Tramontano A et al 1989 Conformation of immunoglobulin hypervariable regions. Nature (Lond) 342:877–883

Fine RM, Wang H, Shenkin PS, Yarmush DL, Levinthal C 1986 Predicting antibody hypervariable loop conformations. II: Minimisation and molecular dynamics of McPC603 from randomly generated loop conformations. Proteins Struct Funct Genet 1:342–362

Wilson IA, Stanfield RL, Rini JM et al 1991 Structural aspects of antibodies and antibody–antigen complexes. In: Catalytic antibodies. Wiley, Chichester (Ciba Found Symp 159) p 13–39

Approaches to the design of semisynthetic metal-dependent catalytic antibodies

Grace R. Nakayama and Peter G. Schultz

Department of Chemistry, University of California, Berkeley, CA 94720, USA

Abstract. The binding site of the Fab fragment of antibody MOPC315 was chemically modified with cyclam (1,4,8,11-tetraazacyclotetradecane) and 1,10-phenanthroline. In the presence of the metal ions cobalt(II), copper(II), gallium(III), iron(II), nickel(II), zinc(II), these semisynthetic antibodies did not catalyse the hydrolysis of amides or phosphate ester substrates, or the oxidation of a 1-hydroxy-2-alkene or of 3,4-dihydrobenzamide substrates. Hydrolysis of a coumarin ester substrate was catalysed by the Fab derivatized with a cyclam moiety with a k_{cat} of 0.063 min^{-1} and K_m of 11.7 μM; the presence of metal ions was not required for catalysis.

1991 Catalytic antibodies. Wiley, Chichester (Ciba Foundation Symposium 159) p 72–90

Antibodies have been shown to catalyse a wide variety of transformations, including pericyclic, photochemical, ligand substitution and redox reactions (Schultz et al 1990, Shokat & Schultz 1990, Schultz 1989a,b, Lerner & Benkovic 1988). A number of different strategies have been used to generate catalytic antibodies, including the use of antibodies to stabilize transition states, bind cofactors, and act as entropic traps. Another approach involves the direct introduction of synthetic or natural catalytic groups into antibody combining sites, either chemically or genetically (Pollack & Schultz 1989, Baldwin & Schultz 1989, Jackson et al 1991). For example, the combining site of the antibody MOPC315 has been selectively modified with an imidazole group by affinity labelling techniques to produce a semisynthetic antibody that catalyses the hydrolysis of 2,4-dinitrophenyl (DNP)-containing coumarin esters (Pollack & Schultz 1989). Site-directed mutagenesis has been used to introduce a catalytic imidazole into the active site of the antibodies MOPC315 (Baldwin & Schultz 1989) and S107 (Jackson et al 1991) to create hydrolytic antibodies. In this paper, we describe the derivatization of an antibody combining site with metal ion-binding cofactors in order to generate semisynthetic antibodies with redox and hydrolytic functions.

A number of semisynthetic proteins with novel properties have been characterized. Kaiser and co-workers have derivatized papain with flavin analogues to create a redox-active flavopapain (Kaiser & Lawrence 1984). Derivatization of the non-specific phosphodiesterase, staphylococcal nuclease, with oligonucleotides or peptides affords semisynthetic nucleases that cleave duplex DNA substrates sequence-specifically (Corey et al 1989, Pei et al 1990). Derivatization of sequence-specific DNA-binding proteins with Fe(II)-EDTA or Cu(II)-1,10-phenanthroline can also generate sequence-specific DNA cleaving proteins (Chen & Sigman 1987, Mack & Dervan 1990). Semisynthetic proteins have been used as catalysts for asymmetric hydrogenation reactions (Wilson & Whitesides 1978).

A number of synthetic enzyme models have been made that contain metal ion-binding ligands. Metal-binding β-cyclodextrins have been shown to catalyse reactions such as oxidation (Matsui et al 1976), ester hydrolysis (Breslow & Overman 1970), and CO_2 hydration (Tabushi & Kuroda 1984). Czarnik and co-workers used β-cyclodextrin conjugated with 1,4,7,10-tetraazacyclodecane-cobalt(III) to catalyse the hydrolysis of *p*-nitrophenylacetate (Akkaya & Czarnik 1988). Sigman and co-workers have shown that zinc(II) catalyses the reduction of 1,10-phenanthroline-2-carboxaldehyde in the presence of an NADH analogue (Creighton & Sigman 1971). Dervan and co-workers have shown that derivatization of sequence-specific DNA-binding molecules, such as distamycin with Fe(II)-EDTA, generates sequence-specific DNA oxidizing agents (Taylor et al 1984). Similarly, derivatization of protein-binding ligands with redox-active metal ions affords bifunctional molecules capable of selectively cleaving proteins (Hoyer et al 1990, Schepartz & Cuenoud 1990).

Design and synthesis

We chose to introduce chemically the cyclam (1,4,8,11-tetraazacyclotetradecane) and 1,10-phenanthroline moieties (Fig. 1) into the binding site of MOPC315 to generate catalytic semisynthetic antibodies. Saturated tetraaza and 1,10-phenanthroline compounds bind a number of metal ions with high association constants (Perrin 1979) and have been used successfully in a number of enzyme model studies. By varying the metal ions, it should be possible to obtain semisynthetic antibodies that catalyse oxidation reactions or the hydrolysis of esters, amides and phosphodiesters.

The antibody MOPC315 binds DNP-containing ligands with association constants ranging from 5×10^4 to $1 \times 10^8\ M^{-1}$ (Haselkorn et al 1974). Lys-52, which is located proximal to the combining site of MOPC315, has previously been modified selectively with a unique thiol through the use of cleavable affinity labels (Pollack et al 1988, 1989). This thiol acts as a unique handle for the

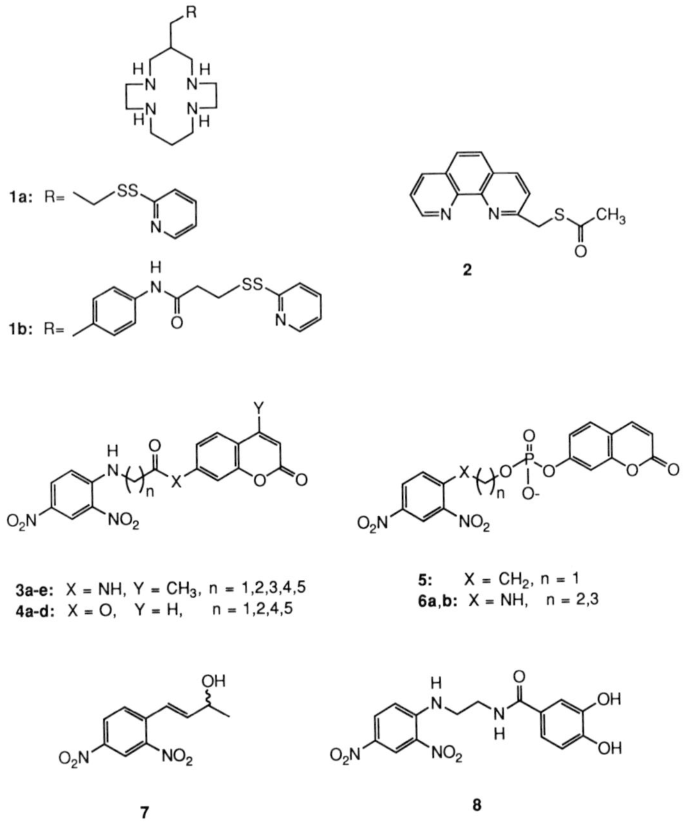

3a-e: X = NH, Y = CH_3, n = 1,2,3,4,5
4a-d: X = O, Y = H, n = 1,2,4,5

5: X = CH_2, n = 1
6a,b: X = NH, n = 2,3

FIG. 1. Structures of cyclams **1a–b** and 1,10-phenanthroline **2** as well as substrates for the hydrolysis (amides **3a–e**, esters **4a–d**, phosphodiesters, **5**, **6a–b**) and oxidation **7,8** reactions.

subsequent introduction of catalytic groups into the antibody combining site. Catalytic groups were introduced into the Fab fragment of MOPC315 by formation of a disulphide bond between the antibody and a thiol-containing catalytic functionality (Fig. 2). Cross-linking was carried out either by reacting the thiopyridylated Fab with the thiolated catalyst, or by treating the free thiol-containing Fab with a catalytic group which contains an S-2-thiopyridyl disulphide.

Two thiol-containing cyclam chelators **1a** and **1b** were synthesized (Figs. 1 and 3) with linkers of various lengths and rigidities between the cyclam ring and the thiol moiety. The cyclam ring in both systems was prepared by a modification of the procedure of Tabushi et al (1977). All intermediates were derivatized with *t*-butyl carbamate (BOC) groups to facilitate purification by silica gel chromatography. The synthesis of cyclam **1a** involved removal of the

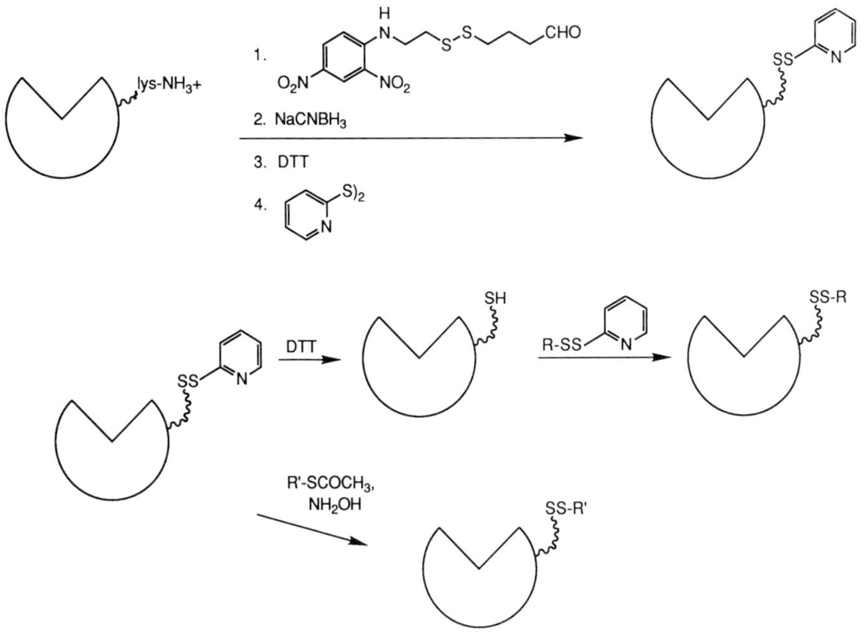

FIG. 2. Modification of the Fab fragment of MOPC315 to generate semisynthetic antibodies.

4-methylbenzyl protecting group by a Birch reduction (maintaining the thiol in reduced form with DL-dithiothreitol [DTT]) followed by treatment with 2,2′-dithiodipyridine to afford the activated S-2-thiopyridyl disulphide. For cyclam **1b**, the nitro group was reduced by hydrogenation and then treated with the symmetric anhydride of S-2-thiopyridyl-3-mercaptopropionic acid to give the activated S-2-thiopyridyl disulphide. The thiolated Lys-52 derivative of the Fab fragment of MOPC315 was treated with 10 equivalents of cyclam **1a** or **1b** in 0.1 M sodium phosphate between pH 5–8 to form the mixed disulphide. In both cases, the reaction was complete within 30 minutes with quantitative release of 2-thiopyridone (determined spectrophotometrically by monitoring the release of 2-thiopyridone at 343 nm [$\Delta\epsilon = 7060\ M^{-1}\ cm^{-1}$]). The semisynthetic antibodies were concentrated by vacuum dialysis and dialysed exhaustively against assay buffer. The desired metal ion was added before assays of catalytic activity.

A thiol group was incorporated into the 1,10-phenanthroline-containing chelator **2** by the synthetic route outlined in Fig. 4. Treatment of 1,10-phenanthroline with methyl lithium, followed by oxidation with potassium permanganate afforded 2-methyl-1,10-phenanthroline. The methyl group was then converted to the corresponding aldehyde by treatment with selenium dioxide

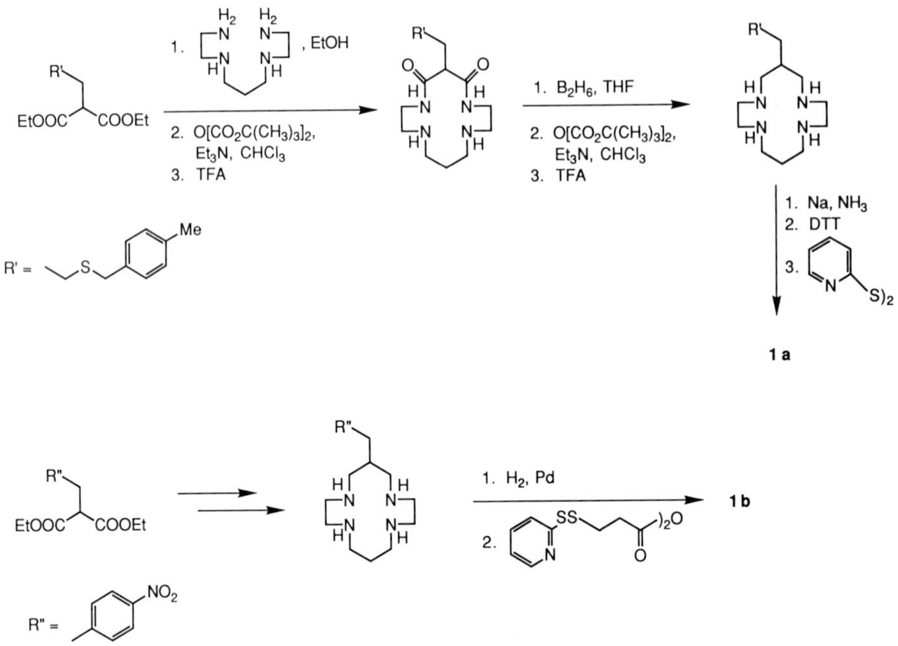

FIG. 3. Syntheses of cyclams **1a–b**.

and subsequently reduced to the alcohol with sodium borohydride. Bromination with hydrobromic acid followed by treatment with potassium thiolacetate afforded the thiol ester of chelator **2**. The free thiol was generated *in situ* by treatment with hydroxylamine in 0.1 M sodium phosphate, pH 7.5 buffer, and then directly reacted with S-2-thiopyridylated Fab to form the mixed disulphide quantitatively. The derivatized antibody was concentrated by vacuum dialysis and dialysed against assay buffer before use. The desired metal ion was added prior to assays of catalytic activity.

Catalytic assays

The semisynthetic antibodies were assayed for their ability to catalyse the hydrolysis of DNP-containing amide **3a–e**, ester **4a–d**, and phosphodiester **5**, **6a–b** substrates. In order to have a sensitive straightforward assay for catalytic activity, we incorporated 7-amino-4-methyl or 7-hydroxycoumarin leaving groups into each substrate. The amide and ester substrates were synthesized by reacting the mixed anhydride of the appropriate N-DNP-amino acid with 7-amino-4-methylcoumarin or 7-hydroxycoumarin, respectively. The phosphodiester substrates were prepared by reacting the appropriate DNP-alcohol with

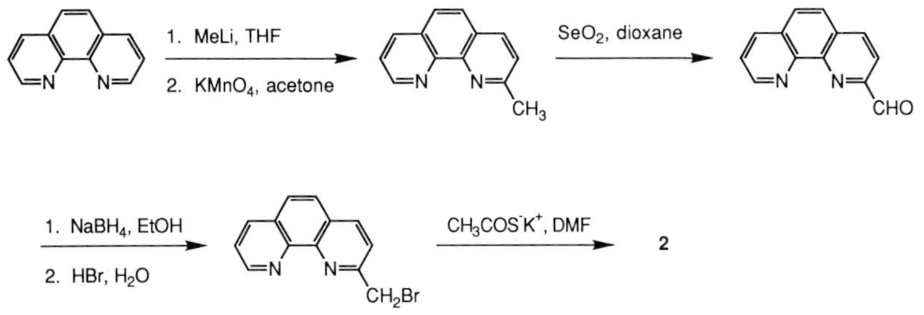

FIG. 4. Synthesis of 1,10-phenanthroline **2**.

the phosphorus oxychloride adduct of 7-hydroxycoumarin. A number of homologous substrates were synthesized in each series to examine the effect of substrate structure on catalytic activity. It has been shown (Baldwin & Schultz 1989, Pollack & Schultz 1989) that variations in substrate structure can have a large effect on the rate of hydrolysis of DNP-esters by imidazole-containing MOPC315 antibodies.

Two substrates were also prepared for oxidation experiments: an allylic alcohol **7** which can be oxidized to the corresponding vinyl ketone, and a 3,4-dihydroxybenzamide derivative **8** which can be oxidized to the quinone or ring-opened diacid (Sheldon & Kochi 1981). Substrate **7** was synthesized by reduction of the aldol product formed from 2,4-dinitrobenzaldehyde and acetone with sodium borohydride in the presence of cerium(III) chloride. Substrate **8** was prepared by coupling the acid chloride of *t*-butyldimethylsilyl-protected 3,4-dihydroxybenzoic acid with N-DNP ethylene diamine.

The amide **3a–e**, ester **4a–d**, and phosphodiester **5**, **6a–b** substrates were assayed for hydrolysis by the semisynthetic antibody in the presence of a variety of metal ions: cobalt(II) chloride, copper(II) chloride, gallium(II) nitrate, iron(II) sulphate, nickel(II) chloride and zinc(II) chloride in 0.1 M triethylamine acetate buffers at pH 6.0, 7.0 and 8.0 at 16 °C. In the amide hydrolysis experiments, the release of 7-amino-4-methylcoumarin was monitored by exciting at 380 nm and measuring emission at 440 nm; in the ester and phosphodiester hydrolysis experiments, the release of 7-hydroxycoumarin was monitored by exciting at 355 nm and measuring emission at 455 nm. Antibody (0.5 μM) was preincubated with 5 μM of the metal salt in reaction buffer for 1–2 minutes, then 5–50 μM of the coumarin substrate was added. Although we did not observe catalysis by any of the semisynthetic antibodies for the amide **3a–e** or phosphodiester **5**, **6a–b** substrates, the antibody derivatized with cyclam **1a** did show hydrolytic activity towards ester substrates **4b–d**. Further experiments, however, showed that the presence of metal ions was not required for the catalytic activity of this semisynthetic antibody. Unmodified MOPC315 also catalysed the hydrolysis of substrates **4c** and **4d** at 16 °C in a buffer containing 0.1 M triethylamine acetate at pH 7.0. Thus, the Fab of MOPC315 derivatized with cyclam **1b** showed enhanced catalytic activity over unmodified Fab only for the hydrolysis of ester **4b**.

The oxidation of substrates **7** and **8** (50 μM) was attempted with 2 μM semisynthetic antibody, in the presence of 50 μM ascorbate and 10 μM iron(II) sulphate or copper(II) chloride at 16 °C in 50 mM boric acid, 50 mM NaCl, pH 7.0 buffer. All samples were degassed with argon before addition of metal ion; oxidation reactions were carried out in an oxygen-containing atmosphere. Reactions were monitored by UV spectroscopy at 268 nm. In all cases, these semisynthetic antibodies failed to show any significant increase in the reaction rate over background.

Thus far, the only reaction catalysed by these semisynthetic antibodies is the hydrolysis of coumarin ester **4b** by the Fab derivatized with cyclam **1a**. Reactions were carried out in the presence and absence of 0.5 μM cyclam **1a**-derivatized antibodies at 16 °C in 50 mM boric acid, 50 mM NaCl, pH 7.0 buffer. The semisynthetic antibody-catalysed reaction rates were corrected by subtracting background rates. An Eadie-Hofstee plot (Eadie 1942, Hofstee 1959) afforded the kinetic constants for this reaction of $k_{cat} = 0.063\ min^{-1}$ and $K_m = 11.7\ \mu M$ (Fig. 5). These values compare with $k_{cat} = 0.052\ min^{-1}$ and $K_m = 2.2\ \mu M$ or hydrolysis of the same N-DNP-coumarin ester substrate **4b** by a semisynthetic imidazole-containing MOPC315 atnibody (Pollack & Schultz 1989). The hydrolysis reaction was inhibited by N-DNP-glycine with an $IC_{50} = 15\ \mu M$. The second-order rate constant in the presence of cyclam was determined at 5 μM **4b** and 5–50 mM cyclam to be $k_{cyclam} = 0.11\ M^{-1} min^{-1}$. The catalytic activity of cyclam **1a**-derivatized Fab is dependent on the length between the DNP and coumarin groups of the ester substrates. Short (**4a**) or long (**4c–d**) tethers result in non-productive interactions of the cyclam with the ester carbonyl.

Discussion

There are several possible mechanisms for hydrolysis of esters by this semisynthetic antibody (Bruice & Benkovic 1966). The cyclam may act as a nucleophile and attack the ester to form a tetrahedral intermediate which is then readily hydrolysed. A second mechanism involves the cyclam acting as a general base to activate a water molecule to attack the ester carbonyl. Alternatively, the cyclam may stabilize a negatively charged transition state by charge complementarity. We are currently conducting a number of mechanistic studies to differentiate between these mechanisms.

A number of arguments can be made for the lack of metal ion-dependent catalytic activity by the cyclam and 1,10-phenanthroline-derivatized semisynthetic antibodies. The addition of a large chelating group near the combining site may block access of substrates to the binding site. However, fluorescence quenching experiments using 2,4-dinitrophenethyl alcohol as the

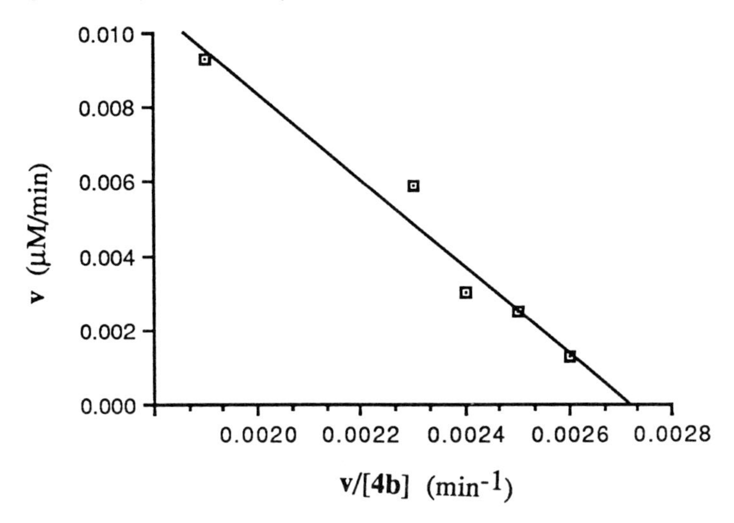

FIG. 5. Eadie–Hofstee plot for the hydrolysis of ester **4b** by cyclam **1a**-derivatized MOPC315 Fab at pH 7.0.

substrate showed that the semisynthetic antibodies had virtually the same binding affinity as unmodified MOPC315 ($K_d = 0.28\ \mu M$). Another possibility is that the cofactors on the derivatized antibodies do not complex metal ions efficiently. Unfortunately, it was not possible to observe the cyclam-metal ion complexes nor mono-1,10-phenanthroline-metal ion complexes spectrophotometrically. Electron paramagnetic resonance studies of the semisynthetic antibodies with copper(II) chloride at −78 °C showed that while copper(II) bound non-specifically to cyclam **1b**-derivatized Fab, it did not form a complex with the chelator. In the case of the 1,10-phenanthroline **2**-derivatized Fab, it was not clear whether the copper(II) complex was formed. Lack of catalytic activity may also result from the fact that the cyclam and 1,10-phenanthroline metal ion complexes are ligated by residues on the surface of MOPC315. In addition, the orientation of the cofactors may not allow productive interactions with bound substrate, or productive interactions may be entropically disfavoured because of the long tethers attaching the cyclam to the antibody.

Further work in this area will involve attaching preformed metal complexes selectively onto the antibodies. Other chelating groups such as porphyrins and EDTA will be tested. Catalytic groups will also be attached to antibodies in which a unique cysteine is introduced by site-directed mutagenesis.

Acknowledgements

We would like to thank S. Pollack for the synthesis of compounds **4a–4d**, and D. Westmoreland for help with the electron paramagnetic resonance experiments. Funding is provided by National Institutes of Health grant # 8R1A124695.

References

Akkaya EU, Czarnik AW 1988 Synthesis and reactivity of cobalt(III) complexes bearing primary- and secondary-side cyclodextrin binding sites. A tale of two CD's. J Am Chem Soc 110:8553–8554

Baldwin EP, Schultz PG 1989 Generation of a catalytic antibody by site-directed mutagenesis. Science (Wash DC) 245:1104–1107

Breslow R, Overman LE 1970 An "artificial enzyme" combining a metal catalytic group and a hydrophobic binding cavity. J Am Chem Soc 92:1075–1078

Bruice TC, Benkovic SJ 1966 Bioorganic mechanisms. Benjamin, New York p 57–58

Chen C-HB, Sigman DS 1987 Chemical conversion of a DNA-binding protein into a site-specific nuclease. Science (Wash DC) 237:1197–1201

Corey DR, Pei D, Schultz PG 1989 Sequence-selective hydrolysis of duplex DNA by an oligonucleotide-directed nuclease. J Am Chem Soc 111:8523–8525

Creighton DJ, Sigman DS 1971 A model for alcohol dehydrogenase. The zinc ion catalyzed reduction of 1,10-phenanthroline-2-carboxaldehyde by *N*-propyl-1,4-dihydronicotinamide. J Am Chem Soc 93:6314–6316

Eadie GS 1942 The inhibition of cholinesterase by physostogmine and prostigmine. J Biol Chem 146:85–93

Haselkorn D, Friedman S, Givol D, Pecht I 1974 Kinetic mapping of the antibody combining site by chemical relaxation spectrometry. Biochemistry 13:2210–2222

Hofstee BHJ 1959 Non-inverted versus inverted plots in enzyme kinetics. Nature (Lond) 184:1296–1298

Hoyer DW, Cho H, Schultz PG 1990 A new strategy for selective protein cleavage. J Am Chem Soc 112:3249–3250

Jackson DY, Prudent JR, Baldwin EP, Schultz PG 1991 A mutagenesis study of a catalytic antibody. Proc Natl Acad Sci USA 88:58–62

Kaiser ET, Lawrence DS 1984 Chemical mutation of enzyme active sites. Science (Wash DC) 226:505–511

Lerner RA, Benkovic SJ 1988 Principles of antibody catalysis. BioEssays 9:107–112

Mack DP, Dervan PB 1990 Nickel-mediated sequence-specific oxidative cleavage of DNA by a designed metalloprotein. J Am Chem Soc 112:4604–4606

Matsui Y, Yokoi T, Mochida K 1976 Catalytic properties of a Cu(II) complex with a modified cyclodextrin. Chem Lett 1037–1040

Pei D, Corey DR, Schultz PG 1990 Site-specific cleavage of duplex DNA by a semi-synthetic nuclease via triple-helix formation. Proc Natl Acad Sci USA 87:9858–9863

Perrin DD 1979 Stability constants of metal ion complexes. Paper B: Organic ligands (IUPAC Chemical Data Series No. 22). Pergamon Press, Oxford, p 872–882

Pollack SJ, Schultz PG 1989 A semisynthetic catalytic antibody. J Am Chem Soc 111:1929–1931

Pollack SJ, Nakayama GR, Schultz PG 1988 Introduction of nucelophiles and spectroscopic probes into antibody combining sites. Science (Wash DC) 242:1038–1040

Pollack SJ, Nakayama GR, Schultz PG 1989 Design of catalytic antibodies. Methods Enzymol 178:551–568

Schepartz A, Cuenoud B 1990 Site-specific cleavage of the protein calmodulin using a trifluoperazine-based affinity reagent. J Am Chem Soc 112:3247–3249

Schultz PG 1989a Catalytic antibodies. Angew Chem Int Ed Engl 28:1283–1295

Schultz PG 1989b Catalytic antibodies. Acc Chem Res 22:287–294

Schultz PG, Lerner RA, Benkovic SJ 1990 Catalytic antibodies. Chem Eng News 68(22):26–40

Sheldon RA, Kochi JK 1981 Metal-catalyzed oxidations of organic compounds. Academic Press, New York, p 231–241

Shokat KM, Schultz PG 1990 Catalytic antibodies. Annu Rev Immunol 8:335–363
Tabushi I, Kuroda Y 1984 Bis(histamino)cyclodextrin-Zn-imidazole complex as an artificial carbonic anhydrase. J Am Chem Soc 106:4580–4584
Tabushi I, Taniguchi Y, Kato H 1977 Preparation of C-alkylated macrocyclic polyamines. Tetrahedron Lett 1049–1052
Taylor JS, Schultz PG, Dervan PB 1984 DNA affinity cleaving sequence-specific cleavage of DNA by distamycin-EDTA-Fe(II) and EDTA-distamycin-Fe(II). Tetrahedron 40:457–465
Wilson ME, Whitesides GM 1978 Conversion of a protein to a homogeneous asymmetric hydrogenation catalyst by site-specific modification with a diphosphinerhodium(I) moiety. J Am Chem Soc 100:306–307

DISCUSSION

Plückthun: Dr Schultz, you mentioned the creation of a hydrolytic antibody, S107, by the introduction of an imidazole using site-directed mutagenesis (Jackson et al 1991). We have done similar experiments with the McPC603 antibody (Glockshuber et al 1991) and get opposite results in terms of the importance of Tyr-33H in McPC603. We find that Tyr-33H is essential in McPC603 and observe a 60-fold loss of phosphorylcholine binding when it is changed to Phe and a 1000-fold loss when it is changed to His, whereas you reported no dramatic difference in the antibody S107 for these substitutions. What is the mechanistic basis for this difference between McPC603 and S107? Clearly, there are three types of phosphorylcholine antibodies that differ in the light chains they use (the McPC603 type, the T15 type and the M167 type), and the problem in modelling the binding sites is that some of the hypervariable loops are different lengths. Also some residues, like the Trp-107H in the McPC603 binding site, are not present in all the monoclonals. If one compares your results on S107 with the results with McPC603, for which the three-dimensional structure is known, one must conclude that the binding is different for these two antibodies.

If we substitute His for Tyr in the antibody T15, we also see a drop in binding, and that molecule should be very homologous to your antibody S107.

Schultz: When we substitute His for Tyr-33H we get about a 10-fold increase in k_{cat}. When we substitute with Phe, we see little effect on binding or catalysis. With a Tyr-33H to Glu mutation we see a significant decrease in the association constant for substrate. The Arg-52H to Lys mutation had little effect on binding or catalysis.

Plückthun: This suggests that the antigen-binding modes of the phosphorylcholine-binding antibodies are much more diverse than anyone has thought.

Schultz: That is probably true, since the antibodies have large binding sites. We have initiated a series of mechanistic experiments on the His mutant to establish the catalytic mechanism.

Benkovic: In the studies involving the oxidation of substituted phenols, is the iron cycling between the II and III oxidation states? Are you reducing it with an auxiliary reagent?

Schultz: We use an auxiliary reductant. We have measured over 200 turnovers by the antibody–haem complex.

Benkovic: Does the porphyrin self-destruct as some of those catalysts do?

Schultz: The antibody–haem complex self-destructs more slowly than free porphyrin. There are two possible explanations: we are preventing dimer formation or we are sequestering hydroxylating species. We have looked spectrophotometrically for a high oxidation state ferryl species but haven't seen one yet.

Benkovic: Have you looked at hydroxylation by this complex?

Schultz: Yes. We have yet to observe any hydroxylation or epoxidation of substrates by the complex.

Schwabacher: We have looked at peroxidase-like oxidations of the same substrates catalysed by antibody–porphyrin complexes (D. Scherübel & M. Weinhouse, unpublished). We find that the rates are slower than with free, soluble porphyrins, although the porphyrin survives longer in the complex.

Benkovic: Peter, some time ago I monitored the ratio of the binding affinity of the transition state analogue to the turnover rate of the substrate, on the grounds that if we had mainly transition state stabilization over background, I would be able to predict the rate acceleration by the antibody over the background rate. For the entropy-driven cyclization reactions, does that ratio fit the rate of antibody acceleration over the spontaneous reaction?

Schultz: The correlation fits perfectly with the porphyrin metallation reaction, where the catalytic mechanism may involve only the distortion of porphyrin. We are trying to correlate k_{cat} with K_m/K_i for several of our antibodies.

Benkovic: The rate ratio is predicted for our cases where we don't appear to have active binding site participation.

Schultz: In cases where we have induced a catalytic group into the antibody combining site, such as the isomerization or elimination reactions, the correlation doesn't hold.

Hansen: In statine there is not only a neutral hydroxyl group but also an extra carbon atom, as it were, in the peptide chain. This extra carbon atom plays a role that I certainly don't understand. Is it a key factor in such protease inhibitors? If so, would such molecules be good haptens for the elicitation of antibodies with protease activity?

Schultz: We have made the hydroxymethyl molecule that corresponds to the transition state; we have not looked at the higher homologue.

Paul: Ascites fluid and bacterial extracts contain many contaminating enzyme activities. When is an antibody preparation pure, in your view? What is your end-point?

Schultz: One purifies the protein to a constant specific activity. For the glycosidases, a protein A column followed by a mono S column will usually remove enzyme contaminants. One can also generate the Fab fragments and assay their catalytic activity. Reproducibility from one ascites preparation to another is an important control. Hapten inhibition, substrate specificity and gel electrophoresis provide additional evidence that one is looking at an antibody-associated activity. Mechanistic experiments and chemical inactivation can also be carried out. Finally, one can spike the antibody with an enzyme and show that the enzyme can be quantitatively removed.

Paul: Are you able to remove the catalytic activity by inhibiting with ligand?

Schultz: Yes. A deoxonojirimycin hapten inhibited a contaminating glycosidase, and peptidyl phosphonates inhibited proteases.

Paul: The ability of the haptenic substrate homologue to inhibit antibody-mediated catalysis in solution is not analogous to affinity chromatography using the substrate itself. Transition state analogues will bind antibodies and contaminating enzymes, perhaps equally well. The hallmark of good antibodies is their ability to bind substrate tightly; most enzymes bind substrates poorly and transiently. So affinity chromatography on a substrate that binds antibody tightly but binds the enzyme weakly should separate the two.

Schultz: I don't view affinity chromatography as a reasonable criterion for purity. In each case the hapten inhibited the contaminating enzymic activity at a fairly low concentration of inhibitor. Consequently, a contaminating enzyme will bind hapten on the column (with excess hapten) and you will not achieve an affinity purification of antibody.

Paul: You need a fairly high affinity for the antibody for the contaminant to stick to the column.

Schultz: Adenosine deaminase binds to a coformyin affinity column.

Paul: What about peptidases? Could peptidase and esterase contaminants from ascites fluids be responsible for the low level esterolysis observed that is attributed to catalysis by antibodies?

Schultz: We have generated what appear to be two or three antibodies that hydrolyse peptide bonds. In each case, the reaction showed saturation kinetics. We also found that only one of five protein A-purified antibodies was catalytically active, which in itself suggests that the activity is not a contaminant. The activity is inhibited by hapten and the reaction showed the expected substrate specificity. However, when the protein was purified further with a mono S column, the catalytic activity was separated from the antibody.

We have purified these antibodies to constant specific activity. Another positive control is that one antibody hydrolyses a D-phenylalanyl ester, whereas enzymes in crude serum do not.

Benkovic: Another important control is an active site titration. If you measure the equivalents of ligand that are bound versus the amount of protein present and demonstrate a 1:1 relationship, you can be confident you are not looking at contaminants.

Lerner: Many of these antibodies have been expressed in all kinds of different systems. You start with an antibody molecule with a certain catalytic activity; somebody takes the same molecule and puts it in a plant and obtains the same activity. Another person makes an Fv fragment from that antibody and finds that same activity.

Paul: There is clearly a weight of circumstantial evidence, but no single experiment appears sufficient to rule out contaminating proteases and esterases. One problem is the low catalytic efficiency of the antibodies—low levels of activity are easily explained by minor contaminants. The problem is minimized if the putative antibody catalyst exhibits high efficiency. The definitive experiment will be to clone the cDNA for the catalytic antibody and show that mutagenesis of critical amino acids results in loss of activity.

Schultz: We have purified the antibodies that catalyse the Claisen rearrangement many times in many different ways, including generating Fabs; we always obtained an antibody with the same catalytic activity, which is not the same as that of chorismate mutase. Two different labs have produced antibodies that do the same thing; Don Hilvert's antibody works enthalpically and ours works entropically. The deuterium isotope effects of the antibody-catalysed reactions are different from those seen with the enzyme.

Hilvert: It is interesting that the catalytic activity you see in your pyridoxal-dependent antibody contrasts with what Raso & Stollar (1975) saw in an analogous system.

Schultz: They used a different conjugation strategy; consequently, the association constants of antibody to pyridoxamine and the amino acid are different. Our linkage was introduced into the side chain of the pyridoxal hydroxymethyl group, to present both elements of the hapten molecule to the immune system and to allow us to build a base into the cofactor rather than the antibody.

Hilvert: That is a very clever idea. But Raso & Stollar saw binding to both components with the antibodies they raised. They didn't see any catalytic advantage for Schiff's base formation, although they saw a small rate acceleration for the pyridoxal-P-catalysed transamination of L-tyrosine.

Schultz: That experiment has to be done carefully. When EDTA is added to the reaction we see no catalytic advantage with the antibody compared to the background reaction. So there is some metal-associated reaction going on that's not specific for the combining site.

Janda: One area where our research overlaps with yours is the transacylation reactions, where we both replace the carbonyl carbon with a β-hydroxylethylene isostere. You have found a number of antibodies that are catalytic to the nitrophenol ester. We have used similar haptenic molecules, without charge, and a similar hydroxyethylene isostere replacement, yet have not found any catalytic antibodies.

It appears that we have conflicting evidence here, but it might be a question of substrate lability. You use nitrophenol as a leaving group; we basically use acetamide phenol.

Schultz: For this reaction, the K_m/K_i ratio isn't very large compared to k_{cat}/k_{uncat}, so other factors are important. It seems that the diversity of the immune system is helping us more than hapten design. That's another advantage of the catalytic antibody field—diversity!

Lerner: When you immunize a mouse, the diversity is not so great as one expects. You obtain many monoclonals that are the same. The people who see this all the time are those who make monoclonal antibodies for mapping of epitope sites and protein molecules; they make many antibodies but find there are only a few distinct molecules. So the numbers are quite small, in some instances.

Hilvert: Even if the immune system has restricted itself enormously, there is still a large population of IgG molecules to draw on. In a typical experiment, only a tiny fraction of the available diversity is examined. When you see that one out of six antibodies that you isolate is catalytic, what does that really mean? Is one out of six statistically significant? I think not.

Green: Dr Schultz, in view of the emphasis generally placed on the importance of the oxyanionic nature of the transition state for hydrolysis, it is surprising that you get catalysis using an antibody raised against a neutral hapten such as a β-(*p*-nitrophenyl)-alcohol. Could there be some small proportion of oxyanion or highly polarized O–H bond as a result of the electron-withdrawing *p*-nitrophenyl group to which the antibody is directed?

Schultz: The nitro group lowers the p*K* of the hydroxyl group by 1.5 p*K* units or less, so it would not be significant.

Green: It does not appear to be significant in the p*K* of the hydroxyl group, but I was thinking about the range of species that may be represented by that hapten and the huge number of antibodies that may be directed to each one. In this way, you might be accessing a small proportion of ionized or highly polarized species when you use the *p*-nitrophenyl hapten that you don't when you use the analogous phenyl hapten.

Schultz: The hydroxymethylene case emphasizes that by searching through tremendous numbers of antibodies one should be able to find great catalytic efficiencies. The λ library technology will help this process.

Lerner: I am also sure that in the world of immunization there have been plenty of 'accidental' transacylations, and if that's a good way to start the immune system, it will happen.

One experiment we tried and never completed for technical reasons was to immunize with carboxyamides in the ground state and then select tetrahedral intermediates, on the grounds that perhaps some antibodies have a serine hydroxyl that attacks the antigen and an intermediate would be formed. In this experiment we wanted to isolate the 'mechanistic' portion of the molecule.

We really wanted to look at only a phosphonate or a phosphonamide, but you can't stop the immune system in its quest for binding energy. You begin to get binding to the 'spinach' (e.g. the benzyl functionality), so you lose the experiment. It would be interesting if one could, in isolation, immunize with the ground state and select for accidental reactivity.

Schultz: We have done a more sophisticated experiment that appears to have worked. We immunized with a hapten designed to affinity label the desired active site amino acid. So if we want a cysteine, we use a reactive hapten that will alkylate it.

Lerner: And that is inescapable for the immune system, because whoever does that, wins the binding energy game.

Paul: It is probably too early to conclude that formation of catalytic antibodies in response to the ground state is entirely accidental.

Lerner: Everything in the immune system is accidental!

Paul: Why should we assume that there is not an active selection procedure in the immune system for the selection of a good catalyst?

Lerner: Enzymes have evolved to react with self and antibodies have evolved to *not* react with self. There is a fundamental difference between the two systems. If you are suggesting that part of the function of the immune system is catalysis, I know of no evidence for that. That's why I use the word 'accidental'. The generally held view is that the earliest defence molecules were not antibodies but the complement proteins, which are highly homologous to snake toxins. Originally, there was this non-specific type of defence where the invaded organism simply threw everything at the offending organism, much as fungi do now with toxins. Then antibodies slipped into the process. The evolution of this binding machine gave directionality to the effectors that already existed. From this point of view, there's no *a priori* reason to presume that the effector functionality depends on catalysis.

Paul: Selection of lymphocytes for clonal expansion is driven by binding of antigen to surface antibody. The higher the binding affinity of the antibody, the greater the probability that the cell will proliferate. There is no direct evidence that the immune system can select for catalytic antibodies; but it is only an assumption that such selection cannot occur.

Lerner: There's a distinction between that which does not happen and that which cannot happen.

Paul: Catalytic antibodies raised against vasoactive intestinal peptide have K_m values in the nanomolar range and they are found circulating in the blood over long periods. This suggests that the immune system 'meant' the antibodies to be there. The antibodies bind the peptide tightly, so selection could be on the basis of binding activity but there is not selection against their catalytic activity (see Paul et al, this volume).

Lerner: You asked if I thought there were accidents in the immune system: I said of course there are accidents, because we deal with chemicals. It is rather like the scorpion that stung the turtle in mid-river, with the result that they both drowned. The turtle asked why it had done that, and the scorpion replied that it's in the genes! In our case, it's in the electrons' orbitals; if you put an amine next to a carbonyl, you may get a Schiff base, you can't help it. Peter Schultz is trying to cause a deliberate accident. If you immunize with a Lewis base, sometimes it will be attacked by a Lewis acid.

Do you see catalysis as one of the main mechanistic ways in which the immune system operates?

Paul: We should not ignore the possibility that catalysis by antibodies may be more than accidental. Catalysis does happen with naturally occurring antibodies that appear to be products of selection on the basis of binding affinity.

Page: We should not exaggerate the difference between catalysis and binding. These are simply extreme manifestations of the same thing. In binding there is a polarization and distortion of electron density to neutralize the inherent charge in each of the binding molecules. In catalysis there is bond-making or bond-breaking, and that's just an extreme form of perturbation of electron density. There is no fundamental philosophical difference between catalysis and binding.

Lerner: I realize that, but the discussion often gets translated into a suggestion that one of the main ways that the antibody system works is by catalysis. I doubt that.

Green: All the catalytic monoclonal antibodies that have been described have been IgG. Has anyone prepared any IgM molecules with catalytic activity?

Schultz: We have looked at a lot of IgM antibodies; reduced and alkylated IgM molecules too. Typically, when we generate antibodies to porphyrins we get a huge IgM response. And when we have a problem with antigenicity, we usually end up with a significant IgM response. We have not obtained a catalytic IgM, so far.

Lerner: We have found some, but the problem of purification makes it difficult to be sure.

Janda: Virtually all our IgM antibodies have some catalytic activity.

Schultz: It may just be that we get an IgM response with problematic haptens and the affinity of the IgM is significantly lower than that of the IgG.

Lerner: That's a cogent point; IgM antibodies appear early in the immune response and tend to be orders of magnitude lower in affinity.

Hansen: Antibodies have been generated that act as *cis/trans* isomerases. One wonders if the substrate could be simply twisted rather than undergoing nucleophilic addition. If one immunized with a cyclohexyl derivative, which didn't have a positively charged nitrogen, one might still see catalysis.

Schultz: We are about to make such a compound.

Hansen: Is it your view that just twisting through 90° will be independent of those charges?

Schultz: No, my guess is that our *cis/trans* isomerase antibody functions primarily by a nucleophilic mechanism, on the basis of chemical modification and pH studies. We are about to test this with a mechanism-based inhibitor.

Schwabacher: If you ran the *cis/trans* isomerization in D_2O, incorporation of deuterium into the product would support the addition/elimination reaction and eliminate a simple twisting mechanism. Very fast isomerization and elimination, sequestered from solvent, would prevent labelling; but if it were seen, it would be definitive.

Hansen: How do you imagine that an antibody molecule could just twist a substrate by 90°? It may happen, but is it a reasonable thing to postulate? And what would occur at the molecular level?

Schultz: It depends on the system. The simplest case in which to test this is in an *o*-disubstituted biphenyl system where there is only one transition state involving rotation around a sigma bond. A planar hapten would be used to mimic the transition state. We are also looking, as you are, at orthogonal amide bonds.

Jencks: There is no reason why you must be able to rotate the bond in the ground state; you simply have to be able to get stabilization of the transition state structure.

Hilvert: Have you tested the possibility that the *cis/trans* isomerase antibody enolizes saturated ketones?

Schultz: Not in this case. In other cases, such as the elimination reaction, we looked for α-deuterium washout but didn't see it.

Blackburn: About 20 years ago, I gave a talk on the distorted amide concept in Cambridge and was criticized by Francis Crick, who said that enzymes are inherently too floppy to distort a substrate in that fashion to achieve catalysis. I left this alone, though later work (e.g. Blackburn et al 1980) supported our ideas on strain catalysis. We are now talking about antibodies as being a lot more rigid than enzymes, which makes me wonder how an antibody would embrace a distorted conformation of an amide in order to achieve catalysis. Bill Jencks used to use the language of 'paying a bit and getting something back'. Would one envisage a step-wise entry of the substrate into the combining site of the antibody, so that the energy penalty for distorting the amide was progressively paid off by binding interactions? It seems to me that if you have to wait for the amide to get orthogonal (carbonyl to nitrogen lone pair), the reaction will never start!

Jenks: I wouldn't say much about anything except the ground state and the transition state of the substrate. If the transition state is tetrahedral, and the enzyme stabilizes it, that is all you need to say. The tetrahedral transition state can be reached, slowly, in the non-enzymic reaction; presumably it is reached more quickly in the enzymic reaction, if the active site is complementary to it. But all we need to think about—I think!—is the transition state.

Hansen: One imagines that the normal uncatalysed mechanism of peptide bond hydrolysis does not involve a 90° twist, but proceeds by the addition of water to form the tetrahedral intermediate. In this strategy, are we choosing too high an energy pathway in the uncatalysed sense, for it to be lowered enough to observe catalysis? The work by Rosen et al (1990) on cyclophilin and FK506 binding protein is crucial to this. You can just say that the peptide bond is sequestered from water. You can also ask how it is that with an uncatalysed prolyl *cis/trans* isomerization, there is no hydrolysis. This raises questions about the transition state and its lifetime. For hydrolysis to occur, I believe the transition state must be stabilized and converted, as it were, to an energy minimum on the potential energy surface. That would be a key difference.

Lerner: What do you think about using external perturbants, like light, not only to perturb bound antigens, but also to alter the nucleophilicity of certain residues?

Schultz: Enzymes do it all the time! Light is cheap, after all: why not use it?

Lerner: When enzymes use light, the general notion is that K_m values are consistent with the sort of light that prevailed during evolution. But when these kinds of energetic conversions are involved, does the enzyme balance itself to the kind of light that is available or could you increase the efficiency by using very high energy light?

Schultz: Could we improve the quantum yield as a function of the wavelength of excitation? That is an interesting idea. We made a catalytic antibody that requires light to cleave thymine dimers. It uses light with a wavelength of 300 nm; the enzyme uses a broad spectrum of light. Excited-state reactions are a whole new area of enzymology.

Lerner: Catalytic antibodies can be used at unphysiological pH levels. They can also catalyse 'unnatural' reactions, like the Diels–Alder reaction. Can we now envisage using 'unnatural' external perturbants?

Schultz: Yes. We are pumping about 80 kcal/mole of energy into the substrate, which would be difficult to do thermally.

References

Blackburn GM, Skaife CJ, Kay IT 1980 Strain effects in acyl transfer-reactions. 5. The kinetics of hydrolysis of benzoquinulidin-2-one: a torsionally distorted amide. J Chem Res (9)294–295

Glockshuber R, Stadlmüller J, Plückthun A 1991 Mapping and modification of an antibody hapten binding site: a site-directed mutagenesis study of McPC603. Biochemistry 30:3049–3055

Jackson DY, Prudent JR, Baldwin EP, Schultz PG 1991 A mutagenesis study of a catalytic antibody. Proc Natl Acad Sci USA 88:58–62

Paul S, Johnson DR, Massey R 1991 Binding and multiple hydrolytic sites in epitopes recognized by catalytic anti-peptide antibodies. In: Catalytic antibodies. Wiley, Chichester (Ciba Found Symp 159) p 156–173

Raso V, Stollar BD 1975 The antibody-enzyme analogy. Comparison of enzymes and antibodies specific for phosphopyridoxyl tyrosine. Biochemistry 14:591–599

Rosen MK, Standaert RF, Galat A, Nakatsuka M, Schreiber SL 1990 Inhibition of FKBP rotomase activity by immunosuppressant FK506: twisted amide surrogate. Science (Wash DC) 248:863–866

Construction of combinatorial antibody expression libraries in *Escherichia coli*

W. Huse

Ixsys Inc, 3550 General Atomics Court, Suite L103, San Diego, CA 92121, USA

Abstract. A λ vector system has been developed for the construction and coexpression of diverse populations of heavy and light chain cDNA sequences. Heavy and light chain sequences within the immunological repertoire are generated by the polymerase chain reaction using primers directed to conserved regions within the variable region framework. Two λ vectors are employed for the independent construction of heavy and light chain cDNA libraries. The libraries are randomly combined to produce a population of λ phage with each containing one heavy and one light chain cDNA sequence. The vectors direct the synthesis and secretion of functional Fab antibody fragments from a dicistronic operon. Libraries of up to 1×10^7 recombinants can be obtained with a diversity approaching *in vivo* estimates. Analysis of the heavy and light chain combinations reveals that the complete antibody repertoire can be generated by subsequent reshuffling of heavy and light chain cDNAs within an initial Fab-producing library. Inherent bias found *in vivo* toward certain heavy and light chain combinations can be virtually eliminated.

1991 Catalytic antibodies. Wiley, Chichester (Ciba Foundation Symposium 159) p 91–102

Monoclonal antibodies have been generated that catalyse a broad range of chemical transformations (Shokat et al 1989, Iverson & Lerner 1989). The rationale is to immunize mice with predicted transition state analogues in the hope that antibodies which bind the analogues specifically will function analogously to known enzymic mechanisms when presented with substrate. Apart from the unpredictable design of appropriate transition state analogues, the probability of finding antibodies where particular amino acid side chains participate in catalysis also depends on the number of different antibodies assayed.

Current methods for generating monoclonal antibodies do not provide an adequate survey of the natural immunological repertoire. For example, in an individual animal there are at least 5–10 000 different B cell clones capable of generating unique antibodies to a relatively small antigen (Schreier 1978, Lawrence et al 1973, Chiller & Weigle 1970). The *in vivo* process of somatic mutation during the generation of antibody diversity additionally generates essentially an unlimited number of unique antibodies. In contrast, owing to the relative inefficiency of the *in vitro* process, the production of monoclonal antibodies generally results in only a few hundred different antibodies.

Successful expression of recombinant antibody molecules in bacterial systems has recently been accomplished (Skerra & Plückthun 1988, Better et al 1988). These prokaryotic systems satisfy all the criteria for assembly of functional antibody fragments. First, synthesis of approximately stoichiometric amounts of heavy and light chains can be accomplished by expressing both chains under the control of a single promoter. Second, transport of both precursor proteins to the periplasmic space occurs efficiently and accurately with correct processing of signal sequences. Third, recombinant antibody fragments are correctly folded into functional domains with correct disulphide bond formation. Fourth, heavy and light chain antibody fragments are self-assembled into functional heterodimers. Additionally, libraries have been constructed which express diverse populations of only heavy (V_H) chain polypeptides (Sastry et al 1989, Ward et al 1989). The antigen-binding affinity of single heavy chains for use as substitutes of Fab fragments is, however, one to two orders of magnitude lower than that of heterodimers. More importantly, heavy chains lose specificity because they tend to be sticky (Ward et al 1989).

The expression of antibody fragment libraries is an important extension of the recombinant expression of unique antibody species. Polymerase chain reaction (PCR) technology has allowed access to the diverse family of antibody genes via conserved sequences in the 5′ and 3′ portions of the variable, the framework and the constant region sequences. Specific amplification of antibody fragments using primers homologous to these conserved sequences allows the construction of libraries containing large populations of heavy and light chain antibody fragments. We report the use of λ phage libraries to generate diverse populations of functional Fab fragments. The system parallels the natural immunological repertoire, and reduces the inherent bias towards a small number of predominant species that is found within an *in vivo* population. Libraries can be constructed and screened for a desired antigen-binding activity in less than two weeks. This efficiency allows the rapid isolation of rare catalytic antibodies in a form suitable for genetic manipulation.

Results and discussion

To obtain a vector system for generating a large number of Fab fragments that could be screened directly, we constructed the expression libraries in modified bacteriophage λ expression vectors. Coexpression libraries are made by first producing separate heavy (V_H) and light (V_L) chain fragment libraries, then randomly combining the two libraries into a single vector population (Huse et al 1989). Each Fab fragment coexpressed from a single vector within the randomly combined population retains all the characteristics exhibited by recombinant antibody fragments expressed singly. Moreover, random combination of both chains serves as a source of diversity and increases the total number of sequences which can be obtained.

The vectors used for expression of V_H and V_L sequences have been previously reported (Huse et al 1989) and are schematically represented in Fig. 1. Populations of V_L sequences are cloned into the λ light chain vector; V_H sequences are cloned into the λ heavy chain vector. The vectors are designed to be asymmetric with respect to the *Not*I and *Eco*RI restriction sites which flank the cloning and expression sequences. The asymmetric placement of these restriction sites in a linear vector allows the V_L-containing portion of the light chain vector to be combined with the V_H-containing portion of the heavy chain vector into a single linear vector for the coexpression of both chains. In addition to the asymmetric restriction sites, each vector includes a leader sequence for the bacterial *pel* B gene (Better et al 1988); a ribosome-binding site at the optimal distance for expression of cloned sequences; and cloning sites for either the V_L or V_H PCR products. The heavy chain vector contains a decapeptide tag at the C-terminus of the expressed V_H fragment for immunoaffinity purification (Field et al 1988).

For the synthesis of V_H and V_L sequences, polyA+ RNA is prepared from the spleen of a single immunized mouse and used in first-strand cDNA synthesis (Aviv & Leder 1972). Resultant DNA–RNA heteroduplexes are used as templates for subsequent amplification by PCR. Primers used for amplification of V_H and V_L sequences are as previously described (Huse et al 1989). V_H amplification is performed in eight separate reactions (Saiki et al 1988) with each reaction containing one of eight 5′ primers and a common 3′ primer. The 5′ primers incorporate an *Xho*I site in the PCR products and the 3′ primer incorporates a *Spe*I restriction site. These sites are used for cloning the amplified products into the heavy chain vector in a predetermined reading frame for expression.

An analogous set of PCR primers is used for the amplification and cloning of mouse κ V_L sequences. These primers incorporate *Sac*I and *Xba*I restriction sites into the PCR products for orientational cloning in the correct reading frame. PCR is performed in independent reactions using five separate sets of primers to ensure unbiased amplification of mRNA sequences. Each reaction contains one of the five 5′ primers and a common 3′ primer. PCR products are pooled and cloned into their respective vectors for the construction of separate V_H and V_L libraries.

The heavy chain library was created from mRNA isolated from the spleen of a 129 G_{IX^+} mouse previously immunized with a *p*-nitrophenyl phosphonamidate transition state antigen (NPN) coupled to keyhole limpet haemocyanin. This primary library contained 1.3×10^6 plaque-forming units (pfu) and at least 80% of the clones express the Fd fragments when assayed by

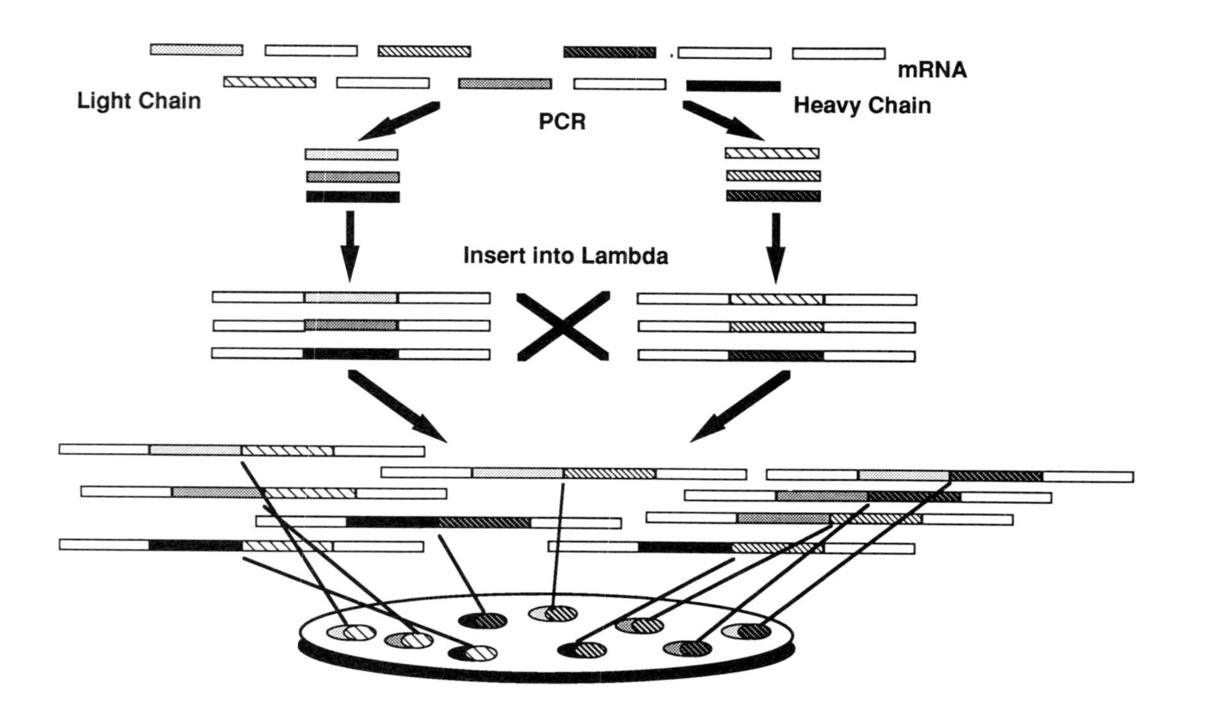

FIG. 1. Scheme for cloning V_H and V_L sequences into λ vectors then crossing the vectors to give coexpression of both chains in a single linear vector.

immunodection of the decapeptide tag. The light chain library was constructed in an identical fashion and shown to contain 2.5×10^6 members. Screening with an antibody to κ chain indicated that 60% of the library expressed light chain inserts.

A single Fab expression library was constructed by ligating the left λ arm containing V_H sequences to the right λ arm containing V_L sequences from each library. To create the library where each vector coexpresses a single heavy and a single light chain-encoding sequence, we performed step-wise restriction of the separate library DNAs to eliminate the λ arms not carrying either encoding sequence. Therefore, the only products which can join together to produce viable phage are the arms containing the heavy and light chain sequences. Light chain library DNA is first digested with *Mlu*I. The ends are dephosphorylated and then digested with *Eco*RI. Preparation of heavy chain library DNA is performed analogously, however the first digestion is with *Hind*III. Products are ligated at their *Eco*RI sites to create a coexpression library.

Reverse immunoscreening of one million phage plaques with labelled NPN coupled to bovine serum albumin (BSA) identified approximately 100 clones which bound to antigen. Antigen–antibody specificity was determined for five of the clones by competition with free unlabelled antigen. The results demonstrated that individual clones could be distinguished on the basis of antigen-binding affinity; the concentration of free antigen for complete inhibition ranged from 10 to 100×10^{-9} M. They also show the production of a recombinant antibody library which compares favourably with the *in vivo* repertoire in terms of size, diversity and antigen-binding affinities.

The mammalian repertoire is typically estimated to contain 10^6 to 10^8 different antigen specificities. Antibody libraries of this size can be obtained using standard recombinant methods and screening techniques to access this number of different molecules in a relatively short time. If more diversity is needed, heavy and light chains within the initial coexpression library or from the original single chain libraries can be systematically shuffled to give larger libraries. Reshuffling of heavy and light chains allows the generation of virtually all possible combinations of V_H and V_L pairs, since the number of individual chains within the library is smaller than the number of possible pairs. V_H and V_L chains can be represented as components of a grid (Fig. 2), with V_H chains along the top row and the V_L chains down the side column. Pairing each chain in a row with each in a column produces a matrix containing all possible combinations (Fig. 2, top). This matrix can be viewed as the total immunological repertoire within an organism. If the initial library does not contain all possible combinations but has each heavy and light chain represented in the population (Fig. 2, middle), then the entire matrix can be generated by reshuffling the chains. The three matrices in Fig. 2 (middle row) represent recombinant antibody populations which might be obtained after the initial construction of the library.

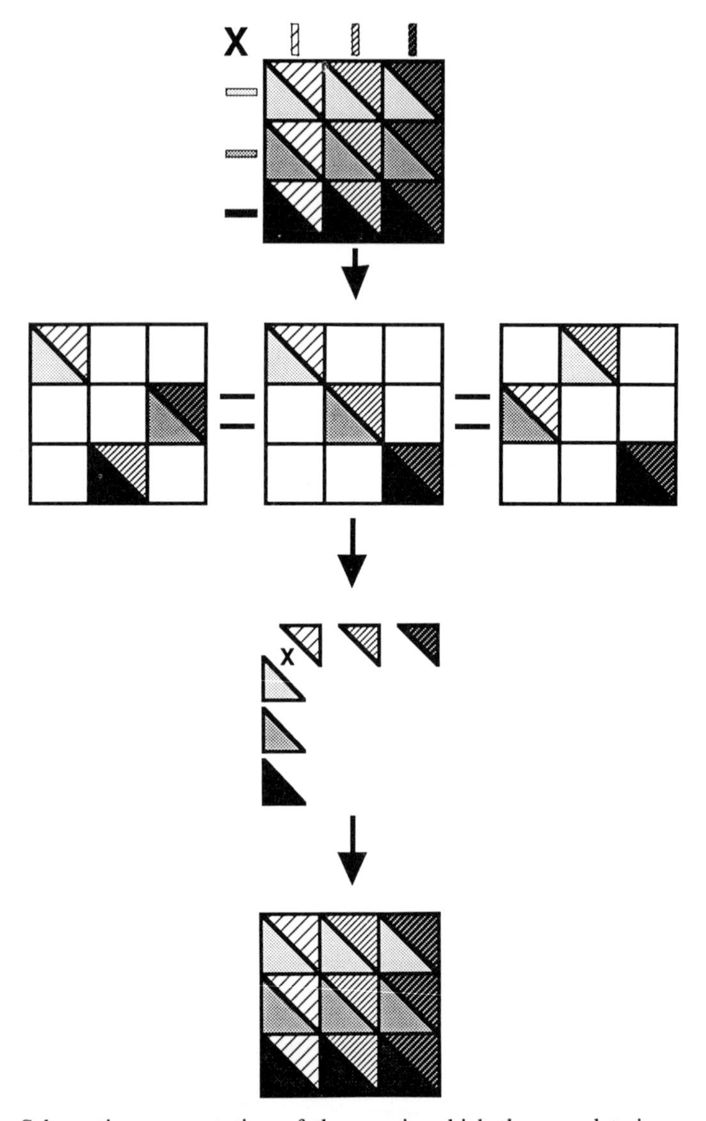

FIG. 2. Schematic representation of the way in which the complete immunological repertoire of an organism can be obtained from individual libraries that contain different subsets of heavy and light chain sequences.

Although they contain different species, the three matrices are equivalent in that each one can be generated from the other by unlinking heavy and light chain pairs and reshuffling. Reshuffling will also generate the complete repertoire of possible species (Fig. 2, bottom). Thus, recombinant libraries can be constructed which faithfully mimic the diversity characteristics of the natural immunological system.

Recombinant antibody libraries also create a relatively unbiased distribution of possible species within the population. Inherent within the immunological repertoire will be prevalence of certain species because of factors such as favoured heavy and light chain re-arrangements and selection of particular heavy and light chain combinations by immunization of the animal. The *in vivo* repertoire has been modelled to account for these factors. Shuffling of the predominant heavy and light chains with the less prevalent chains produces a new population with different characteristics. Diversity within the population does not change, but the frequency of individual species is more equally represented. Unbiased representation of all possible species allows easy access to many more rare antibody species than could previously be obtained.

These results demonstrate that it is now possible to construct and screen several orders of magnitude more clones than using standard immunological techniques. Given the characteristics of recombinant libraries, it may now be possible to isolate antibodies to any desired antigen without immunization of live animals.

References

Aviv H, Leder P 1972 Purification of biologically active globin messenger RNA by chromatography on oligothymidylic acid-cellulose. Proc Natl Acad Sci USA 69:1408–1412

Better M, Chang CP, Robinson RR, Horowitz AH 1988 *Escherichia coli* secretion of an active chimeric antibody fragment. Science (Wash DC) 240:1041–1043

Chiller JM, Weigle WO 1970 Plaque forming cells to 2,4-dinitrophenyl in spleens of immunized and non-immunized mice. Immunochemistry 7:989–992

Field J, Nikawa J, Broek D et al 1988 Purification of a RAS-responsive adenylyl cyclase complex from *Saccharomyces cerevisiae* by use of an epitope addition method. Mol Cell Biol 8:2159–2165

Huse WD, Sastry J, Iverson SA et al 1989 Generation of a large combinatorial library of the immunoglobulin repertoire in phage lambda. Science (Wash DC) 246:1275–1281

Iverson BL, Lerner RA 1989 Sequence-specific peptide cleavage catalysed by an antibody. Science (Wash DC) 243:1184–1188

Lawrence DA, Spiegelberg HL, Weigle WO 1973 2,4-dinitrophenyl receptors on mouse thymus and spleen cells. J Exp Med 137:470–482

Saiki RK, Gelfand DH, Stoffel S et al 1988 Primer-directed enzymatic amplification of DNA with a thermostable DNA polymerase. Science (Wash DC) 239:487–491

Sastry L, Alting-Mees M, Huse WD et al 1989 Cloning of the immunological repertoire in *Escherichia coli* for generation of monoclonal catalytic antibodies: construction of a heavy chain variable region-specific cDNA library. Proc Natl Acad Sci USA 86:5728–5732

Schreier MH 1978 B-cell precursors specific to sheep erythrocytes. Estimation of frequency in a specific helper assay. J Exp Med 148:1612–1619

Shokat KM, Leumann CJ, Sugasawara R, Schultz PG 1989 A new strategy for the generation of catalytic antibodies. Nature (Lond) 338:269–271

Skerra A, Plückthun A 1988 Assembly of a functional immunoglobulin F_V fragment in *Escherichia coli*. Science (Wash DC) 240:1038–1041

Ward ES, Güssow D, Griffiths AD, Jones PT, Winter G 1989 Binding activities of a repertoire of single immunoglobulin variable domains secreted from *Escherichia coli.* Nature (Lond) 341:544–546

DISCUSSION

Schultz: The primers you used have 80% homology with framework 2 sequences from the Kabat sequence collection. When we use those primers we get mis-priming in framework 2.

Huse: Is the homology with framework 2 of the heavy chain or the light chain?

Schultz: Both. In many of our PCR reactions we see two bands on the gel and one must be careful to choose the right band.

Kang: Dennis Burton and I have sequenced a group of mouse and human immunoglobulin variable regions generated by PCR using primers designed to amplify sequences from framework 1. In the case of mouse Fabs that bind NPN–hapten, we have sequenced V_H and V_L for 20 individual clones and have sequence for framework 1 of the heavy and light chains, respectively, for each one. Dennis Burton has observed the same for human V_H and V_L regions.

Huse: We have sequenced 25 randomly chosen clones. All those clones contained framework 1 sequences.

Burton: The upper bands are much higher on the gel. I am surprised if that's true in the sense of putting the decapeptide in frame.

Huse: From the homology of the amino acid sequences, you would expect the decapeptide to be in frame.

Schultz: We also get the decapeptide in frame. I have a compilation of the primers and the matches. In every case there is only a 2 bp mismatch. There is also a third reading frame containing a *lacZα* insert. Would that contribute to the translational inefficiency?

Huse: I wouldn't doubt it. It is difficult to get rid of that site in λ phage. It is very cumbersome to construct anything in λ phage. We are now making some new vectors that will not have that *lacZ* insert.

Shokat: This library is from a mouse immunized with a keyhole limpet haemocyanin (KLH) conjugate. Have you tried screening the light chain library and the heavy chain library separately against a KLH plate?

Huse: We did not test these specific libraries with KLH, but we did test some earlier ones. The problem is that KLH is sticky and when you get spots they don't replicate very well.

Shokat: Could you get heavy chain binding in the absence of light chains, or vice versa?

Kang: We screened the combinatorial library generated from mice immunized with NPN–KLH and found that the KLH sticks to everything. We have no evidence that heavy or light chains alone bind to NPN–bovine serum albumin.

Lerner: Caton & Koprowski (1990) have made a library to influenza virus haemagglutinin using combinatorial libraries. They raised antibodies to the same antigen by hybridoma technology and basically got the same antibodies. They then screened the heavy and light chains individually and did not get any binding.

Paul: That would be a function of binding affinity.

Huse: Binding below a certain affinity (I am not sure what that affinity is) would not be seen using this assay.

Paul: If the assay could detect interactions with weaker affinities, presumably you could pick up binding by single chains.

Huse: The question is, when you do find something that sticks to the heavy chain alone, is that binding specific? If it is not specific, is it predictive of a heavy chain that in combination with a particular light chain will confer specificity? I do not believe the heavy chains that bind non-specifically alone will, in combination with any light chain, necessarily become specific for an antigen.

Mountain: Can you manipulate the stringency of the probing to select for high affinity antibodies?

Huse: I would imagine so, but I don't know. The assay is probably not that sensitive. The concentration of the antibody is not well controlled in those λ phage plaques; different plaques may have very different levels of antibody expression, which will make it hard to separate differences in expression levels from differences in affinity.

Mountain: What's the evidence that you need a bicistronic arrangement of the heavy and light chain gene sequences?

Plückthun: We have used the bicistronic construct, but have also used two different plasmids: the bicistronic construct is the most efficiently repressed. This is important as the induced cells start to lyse. I don't think there is an intrinsic physical reason why the information for both protein chains would have to be on one mRNA molecule.

Mountain: It is not to do with heavy and light chains needing to be secreted in close proximity?

Plückthun: I don't think so. Secretion and folding in each other's presence are what is required, but I don't think that the proteins know whether they were coded on one or two mRNA molecules.

Paul: Dr Huse, the primers you used for amplification by PCR are derived from the amino acid sequences of a limited number of antibodies. Will these primers lead to amplification of every antibody cDNA or nearly every one?

Huse: This method will lead to isolation of the heavy chains that are normally present in the response of a mouse. Dennis Burton and Andreas Plückthun can say more about the diversity of the sequences that are isolated in terms of antibodies to a particular antigen. If we are generating a very diverse set of antibodies to a particular antigen, that would suggest that PCR priming occurs fairly universally on antibody sequences.

Burton: The impression is that you are getting an extremely large, diverse library. That could be true for non-immunized animals, but in an immunized animal, presumably the levels of mRNA from the immunized B cells are much greater, so the repertoire of antibodies generated is limited. Caton & Koprowski (1990) found that one in 200 heavy chains were identical. In the mouse the number of binding sequences is limited: in humans it is greater but still limited. As Richard Lerner has said, the hybridoma technology gives a limited number of sequences.

Plückthun: How do you check the diversity of your plaques?

Huse: From the initial library made with these PCR primers we sequenced 25 randomly chosen clones and they all had different sequences. 20–25% had only a few base pairs different; others had very unrelated sequences.

Hilvert: Gene recombination events are likely to be independent of protein structure, but there might be a structural bias against certain combinations of different heavy chains with different light chains. Is that ever seen in structures of antibodies? Are there certain classes of combinations?

Plückthun: This has been argued for a long time, but there are only a few well designed experiments that address this question. For the different types of phosphorylcholine-binding antibodies, you can obtain cross-association, but you get no antigen binding by any of the cross-associated molecules, only by the original ones. Only in some cases can we understand structurally why the antigen is not bound.

Hilvert: But do you get stable heavy and light chain dimers in the case of the non-binders?

Plückthun: That was what was reported by Hamel et al (1986).

Burton: We have seen the same heavy chain pair with different light chains and still give binding.

Huse: But a heavy chain that binds with more than one light chain still does not bind antigen when combined with most available light chains.

Lerner: If you mutate the framework regions of the antibodies, you get individual specificities. People would argue that in the very high affinity antibodies, not only have you mutated the loops, you have also mutated the dimer interface. Those structures that don't fit any longer drop out. The people who study this effect would argue that private associations occur in less than 5–10% of the very high affinity antibodies.

Hansen: Are any of the antibodies that you raised against the transition state analogue catalytic?

Janda: The problem is that not much protein is produced. Although several different reactions have been catalysed by antibodies, most of the reactions require large amounts of antibody (e.g. 5–20 μM). We have begun to screen some of the antibodies from the library to the nitroanilide phosphonamidate. We began with 20 and have identified one tentatively that cleaves the ester but not the anilide. It must be different from antibody 43C9 because the latter cleaves both the anilide and the ester. We have the DNA sequence of each antibody, so we can see the differences between them.

Hansen: Have you made a library against a transition state analogue that gave a much higher hit rate in the traditional immunization procedure? By traditional immunization, this analogue yielded one catalytic antibody out of 50 you studied; if you are seeing zero out of 20 that could be consistent with the numbers you had before.

Janda: The idea was that by using the library we could find a much greater number of hits.

Schultz: When you screen against that particular hapten for catalysis, won't you run into problems with bacterial enzymes?

Lerner: You can't screen plates for ester and amide hydrolysis. We've tried periodically to screen for substrates in the agar but the bacterial lawn alone often causes the plate to change colour.

Schultz: We're looking at using D esters or amides to do these screens.

Huse: The underlying question is the desirability of doing the assay for catalysis directly without screening for binding initially.

Janda: You would have to have an antibody with a huge turnover number.

Schultz: Or an extremely sensitive assay. We have built into our hapten a fluorescence energy transfer assay which is extremely sensitive.

Lerner: Phage plaques are often screened by enzymic assay, for example the release of halogenated indoles, but β-galactosidase is a very efficient enzyme.

Jencks: Do you have a rough estimate of the affinities of the heavy and light chains compared to that of the intact molecule?

Huse: We don't see anything with the heavy and light chains separately.

Lerner: The problem with heavy chains is they like to stick to things, including light chains and plastic. It is very difficult to study heavy chains alone.

Plückthun: We find exactly the same thing, both the V_H domain alone and the whole heavy chain of the Fab fragment have real problems with solubility. Therefore I am sceptical at least about the generality of using these as antibody substitutes. I don't want to deny that it may work in some cases, but if it's not general, it may not be so useful.

Lerner: The only formal experiment that has been done is that of Andrew Caton. He made a library to a protein molecule, not a small hapten. He divided the library in half and did not see any binding. He could have had, according to the DNA sequences, one in 200 of the heavy chains as proper binders but looking at a million he didn't see any.

Huse: One reason for looking at heavy chains alone was that it was not possible to look at the heavy and light chains together. Hopefully, that is not going to be an issue in the future.

Paul: I am having difficulty relating this discussion to work done in the 1960s and early 1970s. The antibody paratope is a complex of V_H and V_L. Dissociated antibody chains could possess either reduced or no antigen binding activity. It was shown by immunochemical methods that single chains can bind antigen: heavy chains were found to bind better than light chains (Franek & Nezlin 1963, Porter & Weir 1966, Utsumi & Karush 1964). More recently, Greg Winter and co-workers showed that recombinant V_H binds lysozyme (Ward et al 1989). The single chains have reduced binding affinity, but there is little doubt that it is true binding rather than non-specific 'stickiness'.

Huse: The two cases are different. In the first, you take antibodies that bind antigen and ask if their isolated heavy chains have any residual affinity. In the second, you take isolated recombinant heavy chains that stick to an antigen. One would like to know if these are the same population as in the first case. I believe the isolated heavy chains in the second case are probably ones that had residual affinity in the first case plus a bunch of others that are just plain 'sticky' and have nothing to do with specific antibodies. This 'stickiness' could give a background signal that overwhelms the small signal from the residual affinity of heavy chains in the first case.

References

Caton AJ, Koprowski H 1990 Influenza virus hemagglutinin-specific antibodies isolated from a combinatorial expression library are closely related to the immune response donor. Proc Natl Acad Sci USA 87:6450–6454

Franek F, Nezlin RS 1963 Study of the role of different peptide chains of antibodies in the antigen–antibody interaction. Biokhimiya 28:193

Hamel PA, Klein MH, Dorrington KJ 1986 The role of the VL- and VH-segments in the preferential reassociation of immunoglobulin subunits. Mol Immunol 23:503–510

Porter RR, Weir RC 1966 Subunits of immunoglobulins and their relationship to antibody specificity. J Cell Physiol 67(suppl 1):51–64

Utsumi S, Karush F 1964 The subunits of purified rabbit antibody. Biochemistry 39:1329–1342

Ward ES, Gussow D, Griffiths AD, Jones PT, Winter G 1989 Binding activities of a repertoire of single immunoglobulin variable domains secreted from *Escherichia coli*. Nature (Lond) 341:544–546

Catalytic antibodies: contributions from engineering and expression in *Escherichia coli*

Andreas Plückthun and Jörg Stadlmüller

Genzentrum der Universität München, c/o Max-Planck-Institut für Biochemie, Am Klopferspitz, D-8033 Martinsried, Federal Republic of Germany

Abstract. Antibodies have been raised against the transition state of many reactions and shown to catalyse the relevant reaction. Their moderate catalytic efficiencies can be increased by protein engineering, if ways can be found to express the engineered antibody. We have developed a system by which fully functional Fv and Fab fragments can be expressed in *Escherichia coli*. The Fv fragment dissociates at low concentrations; we therefore devised methods to stabilize the fragment. We showed that the Fv fragment of the antibody McPC603, a phosphorylcholine-binding immunoglobulin A, binds the antigen with the same affinity as does the intact antibody isolated from mouse ascites. Phosphorylcholine is an analogue of the transition state for the hydrolysis of choline carboxylate ester. The Fv fragment of McPC603 catalysed this hydrolysis. Mutational analysis of the residues in the binding site of the antibody has shown which are essential for binding and for catalysis, and the importance of charged residues in certain positions. The *E. coli* expression system combined with protein engineering and screening methods will facilitate understanding of enzyme catalysis and the development of new catalytic antibodies.

1991 Catalytic antibodies. Wiley, Chichester (Ciba Foundation Symposium 159) p 103–117

Enzymes use a large repertoire of mechanistic devices to accelerate chemical reactions. One is to exploit the structure of the active site itself. Its shape, hydrophobicity, hydrogen bonds and electrostatic potential not only ensure specific binding of only one or a few substrates, but also contribute in many enzymes to the acceleration of the reaction (for review see Kraut 1988). In enzymes that use catalysis by transition state stabilization, the structure of the active site is more complementary to the transition state of the reaction than to the ground state (Haldane 1930, Pauling 1946). This hypothesis predicts that stable inhibitors which resemble the fleeting transition state should bind more tightly to the enzyme than does the substrate. This prediction has been verified by many experiments (Wolfenden 1976).

Other mechanisms used by enzymes to speed up a reaction include general acid-base catalysis, Lewis acid catalysis, entropic effects, covalent catalysis and the use of coenzymes (Jencks 1969, Walsh 1979, Fersht 1984). Metal ions and coenzymes are the means by which the chemistry of the 20 natural amino acids can be extended. Obviously, the type of reaction to be catalysed determines the mechanisms by which it can be accelerated.

Antibodies share with enzymes the ability to bind chemical compounds rather selectively; this is true for a wide spectrum of different molecules. Moreover, antibodies can be raised specifically to defined compounds. The decisive idea in using antibodies for catalysis was to raise them not to the ground state, but to analogues of the transition state of the reaction, thereby diverting part of the intrinsic binding energy to bringing the substrate closer to the transition state and thus accelerating the reaction (Jencks 1969). The rate acceleration achievable by this strategy is not dramatic, but a whole spectrum of different reactions has now been catalysed (reviewed in e.g. Powell & Hansen 1989, Schultz 1989, Shokat & Schultz 1990). Pioneering investigations (Slobin 1966, Raso & Stollar 1975a,b, Kohen et al 1979, 1980) were hampered by the problem that in polyclonal serum high enough concentrations of the catalytic species are not available, because the antibodies elicited in response to the immunization of an animal make up only a fraction of the total immunoglobulin in the serum. The discovery of monoclonal antibodies (Köhler & Milstein 1975) has made the detailed investigation of antibody-mediated catalysis possible.

The rate accelerations achievable with monoclonal antibodies raised against transition state analogues are comparable to those observed with an enzyme that has been stripped of chemical catalysis. Such an experiment has been performed with subtilisin (Carter & Wells 1988, 1990). The whole catalytic triad (Ser-221, His-64, Asp-32) was replaced by alanine residues. As expected, the rate acceleration was greatly reduced; nevertheless, a k_{cat} of $3 \times 10^{-5}\,s^{-1}$ was observed, which is 2700-fold faster than the rate in solution. This residual catalysis is probably caused by the structural complementarity of the enzyme to the transition state.

Clearly, for high rate accelerations using antibodies, several mechanistic devices have to be combined, just as in 'real' enzymes. Although the immunological repertoire is enormous, it remains doubtful whether immunogens can be devised (except maybe in special cases) that elicit natural antibodies containing, for example, nucleophilic groups or metal-binding sites. The basis of antibody development is selection for tight binding, but this is not necessarily the major determinant of efficient catalysis. Acid-base catalysts, for instance, function optimally if their pK_a is about equal to the pH of the reaction mixture. The immune system, however, selects for *strong* acids and *strong* bases to complement charges in the antigen. Metal-binding sites can probably be created by a purely immunological approach only if a chelating agent is part of the immunogen (Iverson & Lerner 1989), and a metal chelate must then be used as a cofactor. Furthermore, problems will be encountered in screening, if the antibody-producing cells themselves make enzymes with specificities similar to those of the antibody-catalysed reactions for which one is screening.

Why *Escherichia coli*?

These considerations made it desirable to develop methods of protein engineering by which to modify the antibody molecule itself. The techniques for modification of DNA sequences were well established, but no convenient methods were available for expressing the engineered antibody protein. A large amount of work has been done on the expression of antibodies in eukaryotic cells (for reviews, see Brüggemann & Neuberger 1988, Morrison & Oi 1989), but these do not offer the same flexibility or convenience that an *E. coli* expression system does.

Unfortunately, previous attempts to express antibodies and antibody fragments in *E. coli* (Boss et al 1984, Cabilly et al 1984) produced very little functional protein. Most of the protein was obtained as insoluble inclusion bodies. It is, however, the expression in fully functional form that would make the *E. coli* system particularly interesting for antibodies and antibody fragments. Since methods for creating gene libraries in *E. coli* are well established and random mutagenesis of the cloned material is straightforward, catalytic antibodies may then be screened directly on the bacterial colonies by their enzymic activity and/or binding. The relatively simple fermentation of *E. coli* makes large-scale production convenient, which would permit structure determination by both X-ray crystallography (Glockshuber et al 1990b) and NMR spectroscopy.

We have developed a system by which antibody fragments can be expressed in fully functional form in *E. coli* (Skerra & Plückthun 1988, Plückthun & Skerra 1989). This means that small fragments of antibodies can be produced directly without the need for proteolysis. The system was designed for the functional production of Fv fragments (the heterodimer of V_H and V_L; Fig. 1) or Fab fragments (the heterodimer of the whole light chain and the Fd fragment, which consists of the first two domains of the heavy chain). The crucial idea was the simultaneous secretion of both antibody chains into the periplasm, where the oxidizing environment allows the formation of disulphide bonds. Folding of either chain occurs in the presence of the other chain, giving correct assembly. The antibody fragments so produced can be purified by antigen affinity chromatography or immobilized metal ion chromatography (Skerra et al 1991) in a single step. This expression system has been extended to the expression in phage λ of an antibody library generated from an immunized animal (Huse et al 1989).

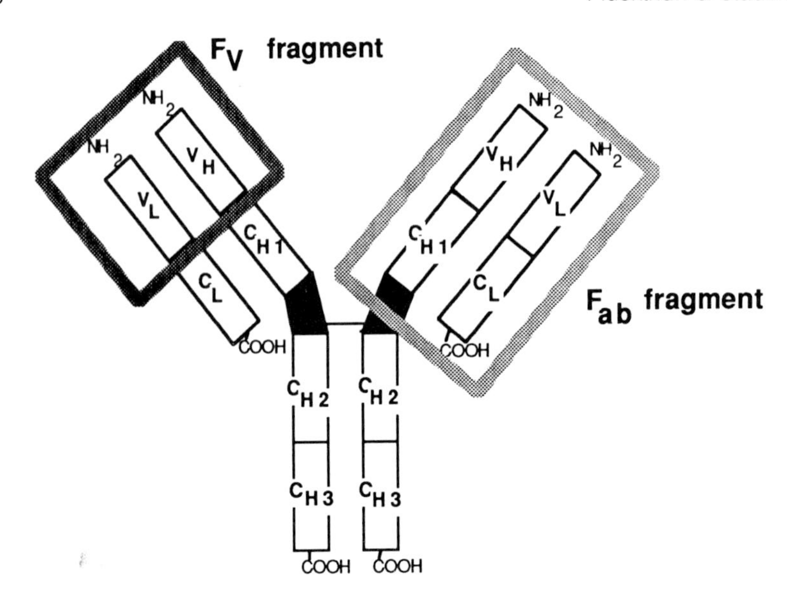

FIG. 1. The Fv and Fab fragments of an antibody, indicated in the schematic drawing of an immunoglobulin of the IgG class. The hinge region is shown in black.

The model system McPC603

Our experiments were carried out with the antibody McPC603, a phosphorylcholine-binding immunoglobulin A from the mouse. This antibody had been particularly well characterized: binding constants and kinetics (Leon & Young 1971, Metzger et al 1971, Young & Leon 1977, Goetze & Richards 1977a,b, 1978), sequences of both chains (Rudikoff & Potter 1974, Rudikoff et al 1981) and, most importantly, the three-dimensional structure are known (Segal et al 1974, Satow et al 1986). It was assumed from the three-dimensional structure that the antibody would catalyse a hydrolytic reaction (see below) (Plückthun et al 1987).

We have shown that the Fv fragment and the Fab fragment produced in *E. coli* have intrinsic hapten-binding constants indistinguishable from those of the whole antibody produced in mouse ascites or those of the proteolytic Fab′ fragment (Skerra & Plückthun 1988, Glockshuber et al 1990a, Skerra et al 1990). Thus our *E. coli* expression system produces fully functional antigen-binding fragments, and the Fv fragment has full antigen-binding capabilities. Before these fragments became conveniently accessible by our expression system, it was not clear whether Fv fragments would be functional (Inbar et al 1972, Sen & Beychok 1986).

At low protein concentrations the Fv fragment dissociates into V_H and V_L (Glockshuber et al 1990a), which leads to complicated hapten-binding behaviour.

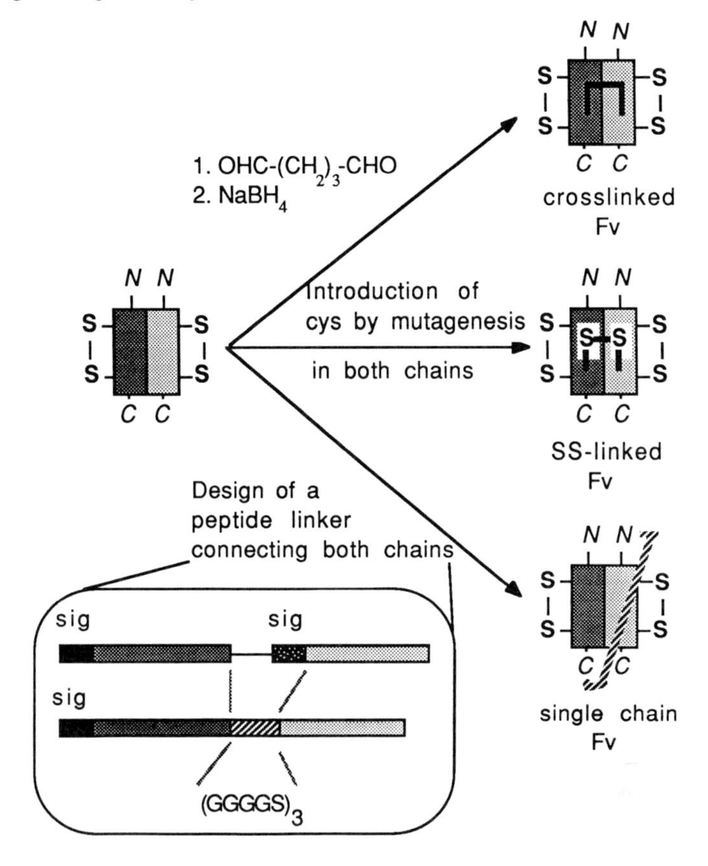

FIG. 2. Strategies designed to stabilize the association of the Fv fragment by linking the domains covalently. The top reaction arrow refers to chemical cross-linking with glutaraldehyde (1) and subsequent reduction (2). The middle arrow refers to stabilizing the Fv fragment by a new interchain disulphide bond, achieved by the introduction of cysteines by mutagenesis into both the heavy (V_H) and light (V_L) chains. The bottom arrow refers to the construction of a secreted single-chain antibody fragment, achieved by site-directed mutagenesis as shown for the genes on the bottom left. The intergenic region and the downstream signal sequence are replaced by a pentadecamer peptide linker $(GGGGS)_3$, which connects the C-terminus of the heavy chain to the N-terminus of the light chain.

We therefore devised three methods for stabilizing the fragment (Fig. 2). The first is chemical cross-linking with glutaraldehyde. The cross-linked Fv fragment shows hapten-binding constants identical to those of the Fab fragment under all conditions tested. Secondly, biochemical cross-linking was achieved by introducing an intermolecular disulphide bond between V_H and V_L at either of two positions (Glockshuber et al 1990a). These disulphide bonds form spontaneously *in vivo* and the resulting fragments have a hapten-binding affinity almost identical to that of the whole antibody. A useful side effect of these modifications is a remarkably enhanced stability to thermal denaturation. Our third strategy was to construct a single-chain Fv fragment in which the two domains are joined covalently by a peptide linker. The single-chain Fv fragments that had been described previously (Bird et al 1988, Huston et al 1988) were expressed as inclusion bodies and refolded *in vitro*. In contrast, we added a signal sequence onto the single-chain Fv fragment and showed that it can then be transported, and normal functionality is obtained with the secreted single-chain fragment. Thus, three different strategies have been developed to solve the problem of chain dissociation in the Fv fragment without increasing its size. It should be noted, however, that the linking of the domains is not necessary, because the fully functional Fv fragment assembles by itself in the periplasm of *E. coli*.

An alternative to using the Fv fragment or the linked Fv fragment is to use the Fab fragment. This does not dissociate at low protein concentrations, because the two constant domains C_L and C_H1 contribute to stability. Interestingly, the yield of functional Fab fragment is consistently lower than that of the Fv fragment under identical conditions. A detailed analysis of this phenomenon (A. Skerra & A. Plückthun, in preparation) has shown that the reason is the lower efficiency of folding and assembly in the periplasm of *E. coli* and not a problem with expression, proteolysis, transport or processing.

Each of these fragments can thus be used for studies of catalysis; the particular applications envisaged will determine which considerations are most important and which fragments should be chosen for the experiment. We have used the Fv fragment in the studies described here, as it can be produced efficiently, and at the high protein concentration required in kinetic measurements dissociation is insignificant.

Catalysis by the recombinant Fv fragment of McPC603

The antibody McPC603 binds phosphorylcholine and esters of phosphorylcholine. Because phosphonates are well-known transition state analogues of peptide hydrolysis and efficient inhibitors of proteases of the metallo- and aspartyl types (Bartlett et al 1987, Kim & Lipscomb 1990), we expected that antibody McPC603 would also stabilize the tetrahedral intermediate of the hydrolysis of a choline-derived carboxylate ester. Similar ideas led Schultz's group to investigate the catalysis of hydrolysis of choline-*p*-nitrophenyl carbonate by the related antibodies T15 and M167 (Pollack et al 1986, Pollack & Schultz 1987). These antibodies are similar to McPC603 but their crystal structures have not yet been determined, and differences in the *length* of loops that contribute residues involved directly in phosphorylcholine binding make model building uncertain.

We investigated the catalysis of hydrolysis of choline-*p*-nitrophenyl carbonate by the Fv fragment of McPC603 expressed in *E. coli*. The kinetic constants

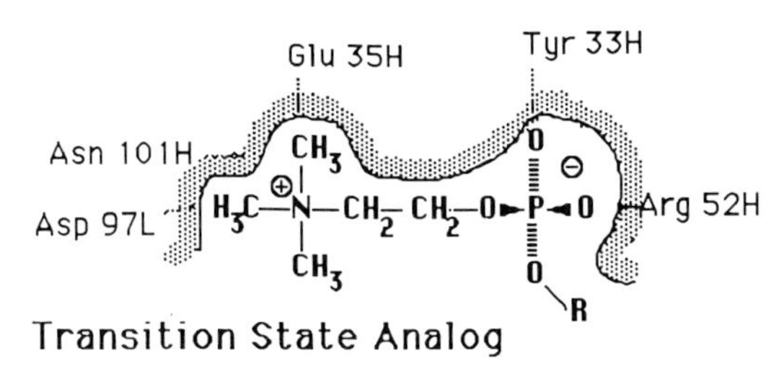

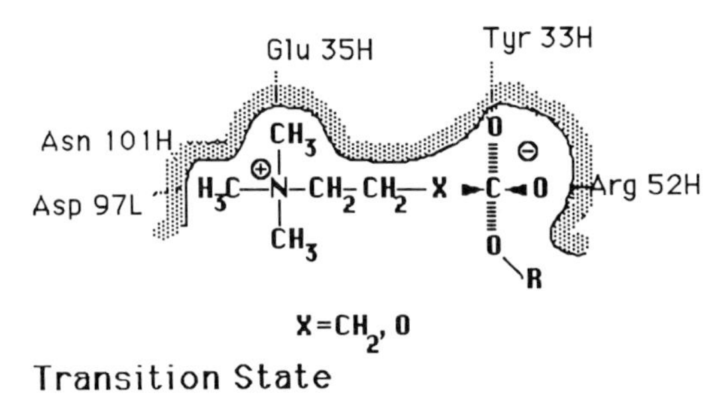

FIG. 3. Scheme for the binding of phosphorylcholine (top) and the transition state (or intermediate) in the hydrolysis of choline-*p*-nitrophenyl carbonate (R = *p*-nitrophenyl) by the active site of the antibody McPC603. The shaded area and marked residues constitute the antibody's binding site. The heavy chain residues Glu-35H, Tyr-33H, Arg-52H and Asn-101H and light chain residue Asp-97L were subjected to mutagenesis.

obtained ($k_{cat} = 0.045\ min^{-1}$ and $K_m = 1.3\ mM$) are lower than those for the T15 and M167 antibodies produced from mouse ascites (Plückthun et al 1990). The hapten-binding constant for phosphorylcholine is identical for the Fv fragment from *E. coli* and the whole antibody, which suggests that the three-dimensional structure of the binding site does not change from antibody to fragment. We must therefore deduce that the difference in hydrolytic rate enhancement between McPC603 and M167 and T15 derives from slight differences in the structures of the binding sites of the different antibodies. Determination of the kinetic parameters of the Fv fragments of T15 and M167 produced in *E. coli* and of their mutants is in progress.

The pH dependence of the hydrolysis of choline-*p*-nitrophenyl carbonate catalysed by T15 and M167 (Pollack et al 1986, Pollack & Schultz 1987) suggests that no acid-base catalysis by the protein takes place. This is consistent with catalysis proceeding merely by transition state stabilization, which may be regarded as a kind of solvent effect. Therefore, the fine details of the binding site appear to determine the magnitude of the rate acceleration.

In a first set of experiments aimed at identifying the residues important for binding, we have mutated those which contact the phosphorylcholine directly (Fig. 3)

(Glockshuber et al 1991). The phosphate must interact with arginine and tyrosine; no other pair of residues tested at positions H33 and H52 functioned as well. The quaternary ammonium group's charge must be compensated by exactly one negative charge in position L97 or H101 (numbering according to the crystal structure of McPC603; Segal et al 1974, Satow et al 1986). No other charge pattern tested led to binding. We are determining the rates of hydrolysis catalysed by these and other mutants to improve our understanding of the relation between binding interactions and transition state stabilization.

Conclusion

Manipulation of the Fv fragment by protein engineering can now be done easily in the *E. coli* expression system that we have developed (Skerra & Plückthun 1988, Plückthun & Skerra 1989, Plückthun et al 1987, 1989). It extends the study of catalytic antibodies in several important ways. First, by using mutants of an antibody with a known structure, further insight into the structural requirements of binding and catalysis can be gained. Second, expression of the antibody in the native state in *E. coli* makes it possible to devise schemes for random mutagenesis, screening and even selection. Third, chemical catalysis in various forms can now be added to the binding site, thus combining various mechanistic devices in the same antibody binding site. Generation of catalytic antibodies by immunization combined with engineering and screening methods should lead to practically useful catalysts and a deeper understanding of enzyme catalysis.

References

Bartlett PA, Marlowe CK, Gianoussis PP, Hanson JE 1987 Phosphorus-containing peptide analogs as peptide inhibitors. Cold Spring Harbor Symp Quant Biol 52:83–90

Bird RE, Hardman KD, Jacobson JW et al 1988 Single-chain antigen-binding proteins. Science (Wash DC) 242:423–426

Boss MA, Kenten JH, Wood CR, Emtage JS 1984 Assembly of functional antibodies from immunoglobulin heavy and light chains synthesized in *E. coli*. Nucleic Acids Res 12:3791–3806

Brüggemann M, Neuberger MS 1988 Novel antibodies by DNA manipulation. Prog Allergy 45:91–105

Cabilly S, Riggs AD, Pande H et al 1984 Generation of antibody activity from immunoglobulin polypeptide chains produced in *Escherichia coli*. Proc Natl Acad Sci USA 81:3273–3277

Carter P, Wells JA 1988 Dissecting the catalytic triad of a serine protease. Nature (Lond) 332:564–568

Carter P, Wells JA 1990 Functional interaction among catalytic residues in subtilisin BPN′. Proteins Struct Funct Genet 7:335–342

Fersht A 1984 Enzyme structure and mechanism. W H Freeman, New York

Glockshuber R, Malia M, Pfitzinger I, Plückthun A 1990a A comparison of strategies to stabilize immunoglobulin F_v fragments. Biochemistry 29:1362–1367

Glockschuber R, Steipe B, Huber R, Plückthun A 1990b Crystallization and preliminary X-ray studies of the V_L domain of the antibody McPC603 produced in *E. coli*. J Mol Biol 213:613–615

Glockshuber R, Stadlmüller J, Plückthun A 1991 Mapping and modification of an antibody hapten binding site: a site-directed mutagenesis study of McPC603. Biochemistry, in press

Goetze AM, Richards JH 1977a Magnetic resonance studies of the binding site interactions between phosphorylcholine and specific mouse myeloma immunoglobulin. Biochemistry 16:228–232

Goetze AM, Richards JH 1977b Structure–function relations in phosphorylcholine-binding mouse myeloma proteins. Proc Natl Acad Sci USA 74:2109–2112

Goetze AM, Richards JH 1978 Molecular studies of subspecificity differences among phosphorylcholine-binding mouse myeloma antibodies using ^{31}P nuclear magnetic resonance. Biochemistry 17:1733–1739

Haldane JBS 1930 Enzymes. Longmans, Green, London; 1965 MIT Press, Cambridge, MA

Huse WD, Sastry L, Iverson SA et al 1989 Generation of a large combinatorial library of the immunoglobulin repertoire in phage lambda. Science (Wash DC) 246:1275–1281

Huston JS, Levinson D, Mudgett-Hunter M et al 1988 Protein engineering of antibody binding sites: recovery of specific activity in an anti-digoxin single-chain F_v analogue produced in *Escherichia coli*. Proc Natl Acad Sci USA 85:5879–5883

Inbar D, Hochman J, Givol D 1972 Localization of antibody-combining sites within the variable portions of heavy and light chains. Proc Natl Acad Sci USA 69:2659–2662

Iverson B, Lerner RA 1989 Sequence-specific peptide cleavage catalyzed by an antibody. Science (Wash DC) 243:1184–1188

Jencks WP 1969 Catalysis in chemistry and enzymology. McGraw-Hill, New York

Kim H, Lipscomb WN 1990 Crystal structure of the complex of carboxypeptidase A with a strongly bound phosphonate in a new crystalline form: comparison with structures of other complexes. Biochemistry 29:5546–5555

Kohen F, Hollander Z, Burd JF, Boguslaski RC 1979 A steroid immunoassay based on antibody-enhanced hydrolysis of a steroid-umbelliferone conjugate. FEBS (Fed Eur Biochem Soc) Lett 100:137–140

Kohen F, Kim JB, Barnard G, Lindner HR 1980 Antibody-enhanced hydrolysis of steroid esters. Biochim Biophys Acta 629:328–337

Köhler G, Milstein C 1975 Continuous cultures of fused cells secreting antibody of predefined specificity. Nature (Lond) 256:495–497

Kraut J 1988 How do enzymes work? Science (Wash DC) 242:533–540

Leon MA, Young NM 1971 Specificity for phosphorylcholine of six murine myeloma proteins reactive with pneumococcus C polysaccharide and β-lipoprotein. Biochemistry 10:1424–1429

Metzger H, Chesebro B, Hadler NM, Lee J, Otchin N 1971 Modification of immunoglobulin combining sites. In: Amos B (ed) Progress in immunology (Proc 1st Int Congr Immunol), Academic Press, New York, p 253–267

Morrison SL, Oi VT 1989 Genetically engineered antibody molecules. Adv Immunol 44:65–92

Pauling L 1946 Molecular architecture and biological reactions. Chem Eng News 24:1375–1377

Plückthun A, Skerra A 1989 Expression of functional antibody F_v and F_{ab} fragments in *E. coli*. Methods Enzymol 178:497–515

Plückthun A, Glockshuber R, Pfitzinger I, Skerra A, Stadlmüller J 1987 Engineering of antibodies with a known three-dimensional structure. Cold Spring Harbor Symp Quant Biol 52:105–112

Plückthun A, Glockshuber R, Skerra A, Stadlmüller J 1990 Properties of F_v and F_{ab} fragments of the antibody McPC603 expressed in *E. coli*. Behring Inst Mitt 87:48–55

Pollack SJ, Schultz PG 1987 Antibody catalysis by transition state stabilization. Cold Spring Harbor Symp Quant Biol 52:97–104
Pollack SJ, Jacobs JW, Schultz PG 1986 Selective chemical catalysis by an antibody. Science (Wash DC) 234:1570–1573
Powell MJ, Hansen DE 1989 Catalytic antibodies—a new direction in enzyme design. Protein Eng 3:69–75
Raso V, Stollar BD 1975a The antibody–enzyme analogy. Characterization of antibodies to phosphopyridoxyltyrosine derivatives. Biochemistry 14:584–591
Raso V, Stollar BD 1975b The antibody–enzyme analogy. Comparison of enzymes and antibodies specific for phosphopyridoxyltyrosine. Biochemistry 14:591–599
Rudikoff S, Potter M 1974 Variable region sequence of the heavy chain from a phosphorylcholine binding myeloma protein. Biochemistry 13:4033–4038
Rudikoff S, Satow Y, Padlan E, Davies D, Potter M 1981 Kappa chain structure from a crystallized murine Fab: role of joining segment in hapten binding. Mol Immunol 18:705–711
Satow Y, Cohen GH, Padlan A, Davies DR 1986 Phosphocholine binding immunoglobulin Fab McPC603: an X-ray diffraction study at 2.7 Å. J Mol Biol 190:593–604
Schultz PG 1989 Antikörper als Katalysatoren. Angew Chem 101:1336–1348
Segal DM, Padlan EA, Cohen GH, Rudikoff S, Potter M, Davies DR 1974 The three-dimensional structure of a phosphorylcholine-binding mouse immunoglobulin Fab and the nature of the antigen binding site. Proc Natl Acad Sci USA 71:4298–4302
Sen J, Beychok S 1986 Proteolytic dissection of a hapten binding site. Proteins Struct Funct Genet 1:256–262
Shokat KM, Schultz PG 1990 Catalytic antibodies. Annu Rev Immunol 8:335–363
Skerra A, Plückthun A 1988 Assembly of a functional immunoglobulin F_v fragment in *Escherichia coli*. Science (Wash DC) 240:1038–1041
Skerra A, Glockshuber R, Plückthun A 1990 Structural features of the McPC603 Fab fragment not defined by the X-ray structure. FEBS Lett 271:203–206
Skerra A, Pfitzinger I, Plückthun A 1991 The functional expression of antibody Fv fragments in *Escherichia coli*: improved vectors and a generally applicable purification technique. Bio/technology 9:273–278
Slobin LI 1966 Preparation and some properties of antibodies with specificity toward p-nitrophenyl esters. Biochemistry 5:2836–2844
Walsh C 1979 Enzymatic reaction mechanisms. W H Freeman, New York
Wolfenden R 1976 Transition state analog inhibitors and enzyme catalysis. Annu Rev Biophys Bioeng 5:271–306
Young NM, Leon MA 1977 The binding of analogs of phosphorylcholine by the murine myeloma proteins McPC 603, MOPC 167 and S107. Immunochemistry 14:757–761

DISCUSSION

Schultz: When you measured binding to the transition state analogue, did you use nitrophenylphosphorylcholine or just phosphorylcholine?

Plückthun: We used phosphorylcholine, but we are now starting a series of experiments with substituted phenylphosphorylcholines.

Schultz: So you were comparing the binding of a substrate to the binding of a different transition state analogue, one with a dianion?

Plückthun: Yes; we have worried about that, which is why we are doing the experiments with nitrophenyl and other phenylphosphorylcholine derivatives. We have done one experiment in which hapten binding was compared with the binding to an affinity column. The ligand on the column was a diester, and therefore a mono-anion. Those mutants of antibody McPC603 for which we saw no binding to the column bound only very weakly to phosphorylcholine. From the one case we have analysed in more detail, which is a mutant with Tyr-33H changed to His, we estimate that failure to bind to the column means at least 1000-fold weaker binding to phosphorylcholine.

Schultz: It is surprising that substitution of Lys for Arg-52H resulted in an antibody with no binding affinity. We saw little change in a related Lys-52 mutant S107 antibody.

Plückthun: I can only speculate because we do not know the structure. There seem to be alternative binding modes for the hapten that may lead to the different behaviour of some of the mutants as well as of the different natural phosphorylcholine-binding antibodies. It may be that the steric requirements of position H52 are such that only arginine can fulfil them in McPC603.

Benkovic: When you say 'no binding', is that really no binding? What difference in affinity can you measure?

Plückthun: 'No binding' is the absence of binding to the affinity column. We have purified the His mutant independently; it has 1000-fold weaker binding and doesn't bind to the affinity column. The Phe mutant has a 60-fold weaker binding and binds to the affinity column. So 'no binding' must be a reduction of binding of between 60- and 1000-fold.

Benkovic: The value for the strength of the electrostatic interactions in your binding studies should be similar to those measured in enzyme mutagenesis experiments. I was really surprised when you said there was no binding, but if 'no binding' is 60–1000-fold weaker affinity, everything is consistent.

Arata: When you have a series of mutants of this antibody expressed in *E. coli*, how do you judge that each mutant has been properly refolded and has a correct tertiary structure?

Plückthun: The evidence is only circumstantial. We have detected the production of these proteins on Western blots; they are properly transported and processed. Some mutants, such as in the cysteine residues, are not made: those I described are not degraded, thus it is likely that they are folded correctly. We cannot rigorously distinguish whether there's a local structural reason for not binding to the affinity column or whether it's a general influence of the folding. That is what we would like to find out.

Arata: Is it possible to make circular dichroism measurements?

Plückthun: Yes, but I don't think circular dichroism measurements are sufficiently sensitive. It is not likely that there will suddenly be a mutation that transforms the antibody into a helix bundle.

Schultz: With regard to the point that a substrate may assume different orientations in the active site, we've carried out electron spin resonance experiments using spin-labelled haptens with phenyl phosphonate-specific antibodies. The spin label in many cases rotates freely, which suggests that in some of these antibodies the carbonyl group of the substrate can adopt many different orientations. There is some correlation between lack of catalytic activity and rotational freedom of the spin label.

Plückthun: I agree this is a likely interpretation. Clearly, we need to determine the preferred binding orientations.

Lerner: If that phenomenon were also true of enzymes, it would explain some of the requirement for holding tightly to the 'spinach' on the substrate that is not mechanistically interesting.

Jencks: That's entropy loss, which accelerates the reaction and *is* mechanistically interesting.

Hansen: Antibody McPC603 is catalytic for a carbonate substrate but not an ester substrate, is that correct?

Plückthun: We haven't done very careful studies on this question, but I think there is product inhibition by the acid from the ester substrate. We know that the acid product binds almost as well as does the phosphorylcholine and presumably there is just a single turnover.

Jencks: Can you make any generalizations about the interaction of the substrate with the binding site from your work in this system?

Plückthun: Binding is really a combination of shape and electrostatics, and shape really means hydrophobic interactions. It seems that even such a small substrate uses hydrophobic forces and that these hydrophobic interactions are essential. The real challenge is to understand this well enough to manipulate it constructively, and this is more difficult than some people believe.

Lerner: If you have a serine protease and study it to its finest detail, you probably understand how all serine proteases work. But if you have a diversity system with a large number of members, like the immune system, and study one member (antibody) to its finest detail, will you know anything about the next one?

Plückthun: I will know what I am aiming for, what my next transition state analogue should look like.

Lerner: Why would you know that?

Huse: If the transition state analogue is very similar to the previous one, then you will have some information. I agree that if it is different, you will have no useful information.

Plückthun: I doubt whether the purely immunological approach can make chymotrypsin.

Huse: So you think 'catalytic antibodies' is an oxymoron?

Plückthun: I think we will be able to obtain decent catalytic activities, but we won't obtain the equivalent of enzymes without using protein engineering.

Schultz: I am willing to bet a case of champagne that we can make a catalytic antibody that functions as well or better than the enzyme chorismate mutase within two years.

Lerner: I am willing to bet another case that the combination of factors, including metals, will approach the efficiency of chymotrypsin. For certain substrates, the nitroanilide catalytic antibody, NPN43C9, that Steve Benkovic has studied (this volume) is only a few orders of magnitude less efficient than chymotrypsin.

Plückthun: The example of metals illustrates exactly what I am saying—metals are put in by engineering. You can't immunize an animal such that the histidines are generated where you want them.

Lerner: That depends on the numbers. Secondly, you must immunize an animal to start with to give the substrate specificity. The rate of catalysis, to me, is not the intellectual heart of the matter.

Protein engineers would like, for example, to make lysozyme into trypsin. When you ask how they plan to do that they say, first I am going to take lysozyme and make it bind to an amide. After that I am going to make it catalytic. Using the immune system the first part is solved and that's the hard part. Extraordinarily high substrate specificity and discrimination between substrate and transition state are already programmable within the immune system to some extent. We don't approach the problem like a blind watchmaker who can't see the solution, but rather like a myopic watchmaker! We see the solution dimly and then use the immune system to refine it.

Plückthun: All of this would benefit from knowing, at least in a few cases, which structures of catalytic sites have worked and which haven't worked.

Lerner: No doubt, but we should not be too preoccupied with rate accelerations.

Jencks: Part of the problem is that we still cannot account very clearly for the catalytic activity of an ordinary enzyme. The distinction here is between complementary binding, which can be done quite well with antibody systems, and chemistry, which involves covalent bonds, general acid-base catalysis and so forth. The chemistry does not arise directly from the complementarity.

Lerner: But there is a certain inability to discriminate between binding and chemistry. If you protonate a leaving amine in an amide bond hydrolysis in a rate-limiting step, is that chemistry or binding? It is both.

Jencks: Chemistry usually requires different groups than does binding, although they can be the same.

Lerner: I do not dispute that we need a knowledge base. There are several X-ray structures of antigen–antibody complexes. Often, if there is a guanidinium in the antigen, there will be a carboxylate in the antibody.

Plückthun: From the organic chemist's point of view, if you use a guanidinium to bind a phosphate group, you would prefer bidentate binding. This is not what happens in the antibody McPC603; that is very clear from the electron density. If one simply approaches this problem using general principles without having any detailed information, one is not going to make much progress.

Jencks: It's important to get some values for the energies of these interactions.

Lerner: The problem that Ian Wilson has, for example, is which crystal structure to solve. It is very different generating information in a diversity system and in a single protein which represents an evolutionary end-point.

Schultz: We have studied the phosphorylcholine-specific antibody, S107, and Andreas Plückthun has examined McPC603—they are quite homologous and yet we see completely different results with the Lys to Arg and Tyr to His mutations. So even for a family of related antibodies you can't generalize.

Wilson: For McPC603, one might assume that the charged groups were in approximately the same position in the antibody combining site, but this may not be the case.

Plückthun: The key point is that we don't want to make this assumption; we want to make observations. We need to obtain a different crystal form of the Fv fragment and try to get a structure of the mutants, as we have done for the light chains. For the light chains the structure determination has been very efficient—we could get a structure, including mutagenesis, purification, crystallization and refinement, within a few weeks.

Wilson: In antibody HyHEL5, if you substitute Arg-68 with Lys, you lose three orders of magnitude of binding avidity (Smith-Gill et al 1982).

Lerner: If you look at Allen Edmundson's light chain dimers, he claims that hydrophobic molecules go into the pocket and the flexible side chains are moved to wherever they need to be to accommodate binding.

Plückthun: We see this here as well. We have tried to model the light chain dimer based on the Bence-Jones protein REI, which is highly homologous to the light chain of McPC603. For the individual domains, the model was very good. For the orientation of the two domains with respect to each other, there was a small difference in the angle, even though the interfaces are almost identical. That is another variable that is very poorly understood.

Paul: Concerning the issue of diversity, antibodies with different amino acid sequences in their hypervariable segments can express similar binding specificities. Is it likely that catalytic mechanisms used by antibodies would show similar redundancy?

Huse: Are you asking whether there are more solutions to a particular catalytic reaction than to a binding reaction?

Paul: Enzymes appear to use a restricted set of mechanisms, and the structural demands for catalysis may be more stringent than those for binding. My intuitive feeling is that there would be fewer solutions for catalysis.

Plückthun: One has to define what one means by a solution. From a purely physical point of view, the solution would be an electric field around the substrate molecule with protons, point charges and field vectors in certain directions. In that case there are probably only very few solutions to catalytic mechanisms.

But there are many molecular arrangements or architectures that could create this electrostatic field with the protons and the hydrophobic interactions in the right place.

Hansen: It would be interesting to study carefully a reaction that is catalysed by a number of different antibodies to see how many solutions we have available.

Lerner: Both kinetically and physically. I grew up in the age of immunology when the histocompatibility system was just being discovered. I would go to meetings and people would describe the histocompatibility system molecule by molecule. Finally, one genius stood up and said, what's really interesting is that they are all different. We have to be very careful in the way we think about diversity systems rather than end-point evolutionary systems.

References

Benkovic SJ, Adams J, Janda KD, Lerner RA 1991 A catalytic antibody uses a multistep kinetic sequence. In: Catalytic antibodies. Wiley, Chichester (Ciba Found Symp 159) p 4–12

Smith-Gill SJ, Wilson AC, Potter M, Prager EM, Feldmann RJ, Mainhart CR 1982 Mapping the antigenic epitope for a monoclonal antibody against lysozyme. J Immunol 128:314–322

The generation of antibody combining sites containing catalytic residues

Kevan M. Shokat and Peter G. Schultz

Department of Chemistry, University of California, Berkeley, CA 94720, USA

Abstract. To expand the scope of antibody-catalysed reactions to those involving rate-limiting proton abstraction, such as elimination, isomerization and condensation reactions, we developed a new strategy—hapten charge complementarity. A hapten containing a benzyl ammonium group was used to elicit a specific base, a carboxylate, in the combining site of an antibody that catalysed a β-elimination reaction. This was the first example of the use of a hapten to elicit a specific catalytic residue in an antibody combining site. A variety of kinetic and chemical modification experiments strongly suggest that a specific Asp or Glu residue in the combining site is responsible for catalysis. Preliminary results indicate that in addition to charge–charge complementarity, the nucleophilic reactivity of amino acid residues (Ser, Thr, Lys, Asp, Glu, Cys) in antibodies can be used as a selection tool. Antibodies were raised against a reactive epoxide group to elicit an antibody containing a uniquely reactive carboxylate or thiol group. Antibodies which bind the epoxide do catalyse a β-elimination reaction, indicating the presence of a specific base in the combining site. Antibodies elicited to two closely related haptens do not catalyse the β-elimination reaction.

1991 Catalytic antibodies. Wiley, Chichester (Ciba Foundation Symposium 159) p 118–134

The humoral immune system's ability to generate 10^8 unique antibody combining sites has been exploited for numerous applications in biology and medicine. In the past five years this diversity has also been exploited for the production of highly selective catalysts (Schultz 1989a,b, Schultz et al 1990, Shokat & Schultz 1990). The challenge in generating antibody catalysts is to design appropriate antigens on the basis of our understanding of enzymic catalysis. A number of successful strategies have been developed which fall into two broad categories. The most widely used method exploits the steric and electronic complementarity of an antibody to its corresponding hapten. This approach allows (a) generation of precisely positioned catalytic amino acid side chains in the combining site, (b) stabilization of transition states in reactions, (c) reduction in the entropy of reactions by orienting participating groups in reactive conformations, and (d) the incorporation of cofactor-binding sites into antibody combining sites. The second general approach for generating catalytic antibodies involves direct introduction of catalytic groups into the antibody combining site by selective chemical modification, site-directed mutagenesis, or genetic selections or screens. The development of a general set of 'rules' is essential for optimal exploitation of the diversity of the humoral immune system.

One of the most general rules to emerge from catalytic antibody studies is that electrostatic complementarity can be used to generate antibodies with precisely positioned positively or negatively charged residues in combining sites. Residues selected in this manner have played catalytically essential roles in reactions such as β-elimination (Shokat et al 1989), trityl group cleavage (Iverson et al 1990) and acyl transfer reactions (Jacobs et al 1988, Janda et al 1990, Pollack & Schultz 1987, Tramontano et al 1986). The question then arises, how does one select for amino acids such as Ser, Cys, Thr and His, which play key catalytic roles in the mechanisms of many enzymes? Antibody selection typically depends on high affinity non-covalent interactions between a hapten and the complementary antibody, such as the electrostatic interactions noted above. In this paper we report preliminary results of a new strategy termed covalent antigenicity, which is based on an antibody's ability to add covalently to a chemically reactive hapten. Association of antibody with antigen in this case depends on formation of a covalent bond between hapten and reactive amino acid side chains in the antibody combining site. Consequently, it should be possible to elicit an antibody with a desired amino acid residue by introduction of an affinity label (for example, epoxide, bromoketone or enone) into the hapten that is specific for that residue. Preliminary studies indicate that antibodies elicited by this strategy successfully catalyse a β-elimination reaction.

Charge–charge hapten complementarity

Antibody–hapten charge–charge complementarity was first demonstrated by Pressman and Grossberg (Grossberg & Pressman 1960, Pressman & Siegel 1953). Antibodies elicited to negatively charged haptens were shown to contain specific Arg and Lys residues in the combining sites. Conversely, positively charged ammonium-containing haptens elicited antibodies with Asp and Glu residues in their combining sites. From these studies we reasoned that catalytic antibodies could be generated which contained catalytic groups within bonding distance of substrates by judiciously introducing charged groups into substrate-like haptens.

To test these ideas, we assayed monoclonal antibodies raised against hapten **1** for their ability to catalyse HF elimination from the fluorinated substrate **2**. The hapten contains a positively charged ammonium ion in place of the α-CH_2 group of the substrate (the tertiary nitrogen obviates competing lactam formation during the coupling of **1** to protein carriers for immunization).

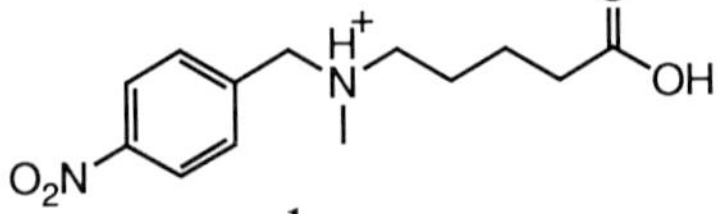

O
H+
N
OH
O_2N
1

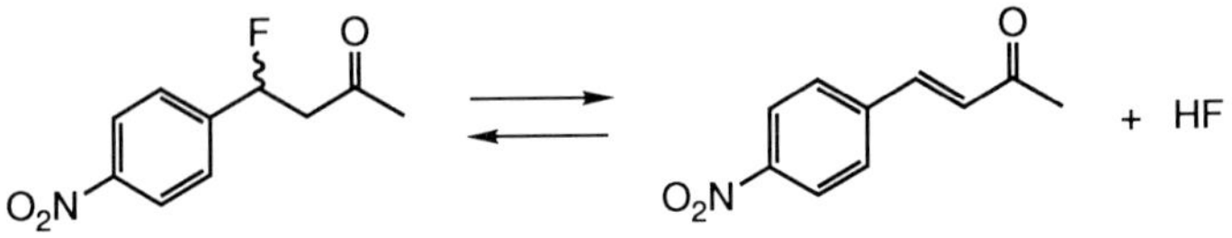

F O
O_2N
2

O
+ HF
O_2N
3

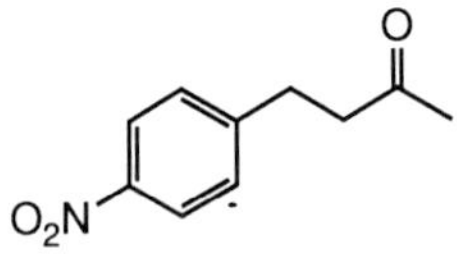

O
O_2N
4

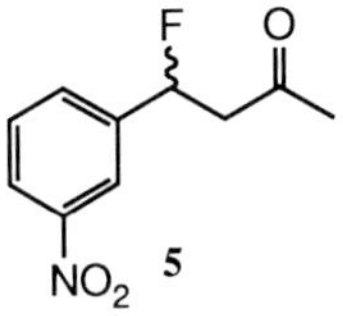

F O
NO_2
5

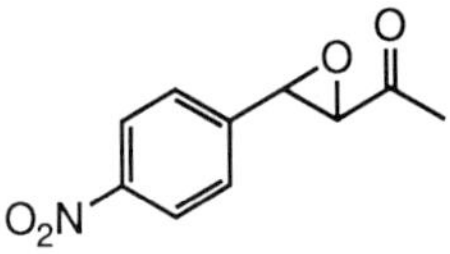

O
O
O_2N
6

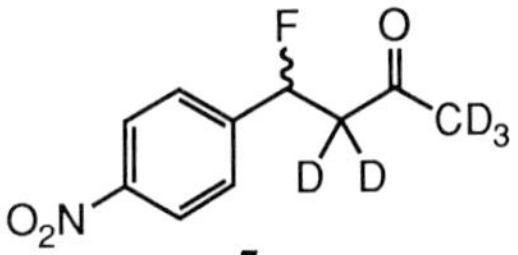

F O
CD_3
D D
O_2N
7

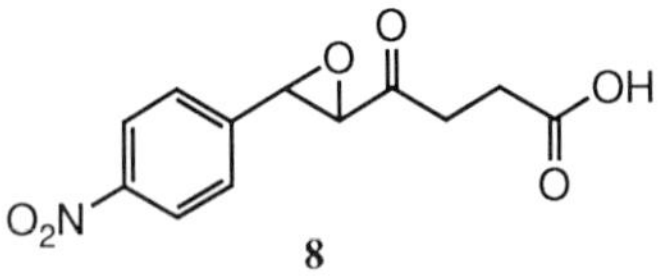

O
O
OH
O
O_2N
8

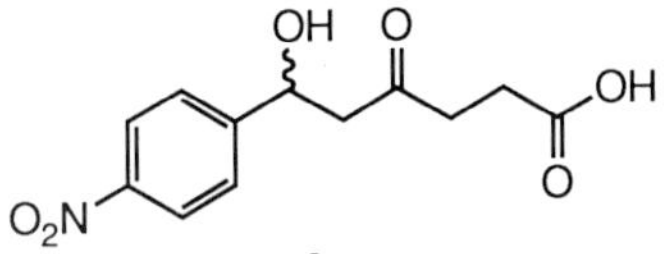

OH O
OH
O
O_2N
9

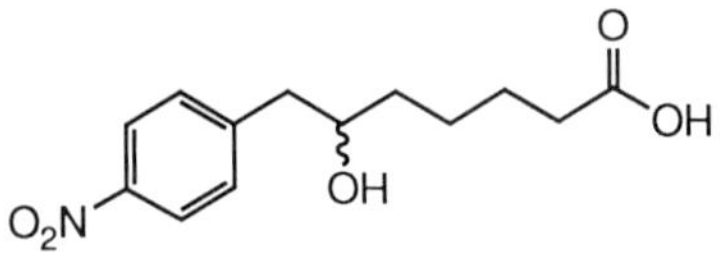

O
OH
OH
O_2N
10

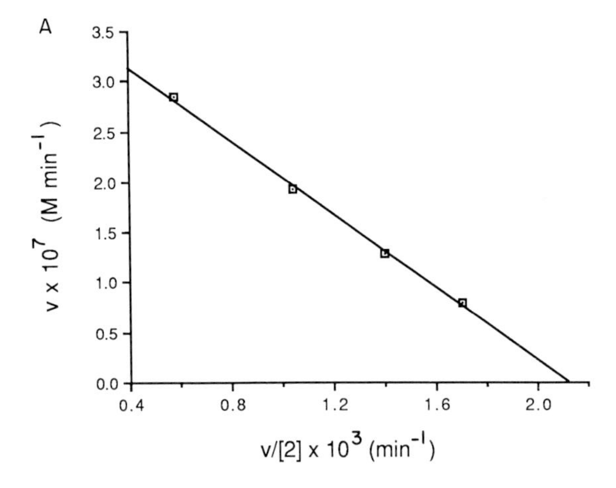

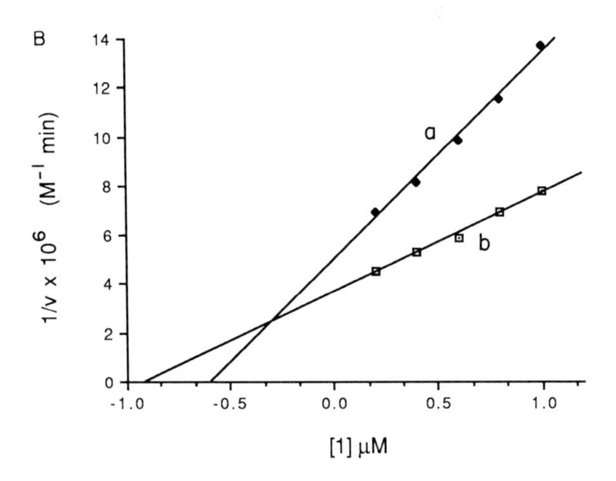

FIG. 1. (A) Eadie–Hofstee plot for the β-elimination conversion of **2** to **3**, catalysed by antibody 43D4-3D12. Velocities were determined spectrophotometrically by measuring the initial absorbance increase at 330 nm. The concentration of 43D4-3D12 was 2 µM as determined by absorbance at 280 nm using E(1 cm, 0.1%) = 1.37 and $M_r = 150\,000$ for immunoglobulin G. 43D4-3D12 was preincubated at 37 °C in 10 mM bis(2-hydroxyethyl)imino-tris-(hydroxymethyl)-methane (bis-Tris), 100 nM NaCl, pH 6.0. The reaction was initiated by adding 10 µM of a stock solution of **2** in CH_3CN to give a final CH_3CN concentration of 2%. (B) Dixon plot of inhibition of 43D4-3D12 by **1**. Antibody concentration, 2.01 µM. Data were obtained at two concentrations of **2**: a, 208 µM; b, 514 µM. Buffer conditions as above.

The hapten and substrate share a recognition element, the *p*-nitrophenyl group, which ensured that the antibody binding affinity for the substrate would be reasonable. Moreover, replacement of hapten by substrate in the antibody combining site should lead to an increase in the basicity of the catalytic carboxylate since a stabilizing salt bridge interaction is lost.

Antibodies were generated against hapten **1**, which at physiological pH exists as a positively charged alkyl ammonium ion (Shokat et al 1989). The hapten was conjugated to the carrier proteins bovine serum albumin and keyhole limpet haemocyanin using the activated *N*-hydroxysuccinimide ester of **1**. Six monoclonal antibodies specific for **1** were produced *in vivo* and isolated by affinity chromatography on protein A-coupled Sepharose 4B, as described previously (Jacobs et al 1988). Of these, four accelerated the β-elimination reaction of **2** to **3** (Fig. 1) and were completely inhibited by the addition of free hapten **1**. This suggests that the catalysis is due to binding of the substrate in the antibody combining site (non-hapten-specific antibodies did not catalyse the elimination reaction). The fact that two-thirds of the antibodies were catalytic demonstrates the general applicability of this strategy for producing catalytic antibodies. One of the four catalytic antibodies, 43D4-3D12, was characterized further.

The reaction catalysed by 43D4-3D12 obeys classical Michaelis–Menten kinetics: a k_{cat} of 0.193 min^{-1} and a K_m of 182 μM for substrate **2** were measured at pH 6.0 (Fig. 2). The catalysed reaction is competitively inhibited by hapten **1** (K_i = 290 nM) and by substrate analogue **4** (K_i = 280 μM), demonstrating that the catalytic activity is associated with binding in the antibody combining site. As expected, the antibody-catalysed reaction is substrate specific, in accordance with the characteristic specificity of antibodies for their ligands (Pressman & Grossberg 1968). The K_m and k_{cat} values for 4-fluoro-4-*m*-nitrophenylbutan-2-one (**5**) are 571 μM and 0.079 min^{-1}, respectively (k_{cat}/K_m = 1.38×10^2 M^{-1} min^{-1}) at pH 6.5, compared with a k_{cat} of 0.304 min^{-1} and K_m of 214 μM (k_{cat}/K_m = 1.42×10^3 M^{-1} min^{-1}) for **2** at this pH.

The identity of the catalytic amino acid side chain in the antibody combining site was investigated using three methods. The pH dependence of k_{cat} shows the classical profile for catalysis attributable to a single titratable group (Fig. 2). The pK_a for the active site residue was 6.2. This is very close to the pK_a for the active site Glu-135 in carboxypeptidase A (pK_a = 6.0), which is known to be responsible for the catalysis of a similar β-elimination reaction (not the physiologically relevant reaction for this enzyme) (Spratt & Kaiser 1984). The maximum k_{cat} achievable by the antibody occurs when the residue responsible for base catalysis is fully deprotonated; from the pH profile determined by plotting k_{cat} versus $k_{cat}[H^+]$ it is 0.458 min^{-1} (Fersht & Renard 1974).

Chemical modification confirmed that an Asp or Glu residue is involved in the elimination reaction. Antibody 43D4-3D12 was treated with diazoacetamide, which reacts almost exclusively with carboxylate groups to form glycolamide

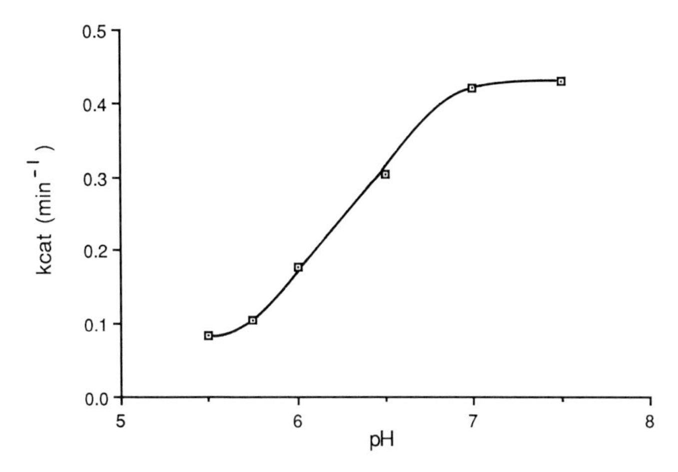

FIG. 2. The k_{cat} versus pH profile for conversion of **2** to **3** catalysed by antibody 43D4-3D12. At each pH the k_{cat} value was obtained from a Lineweaver-Burk plot. The buffer used was 10 mM bis-Tris plus 100 mM NaCl.

ester linkages (Grossberg & Pressman 1960). The antibody 43D4-3D12 retains only 23% of its catalytic activity after treatment with diazoacetamide; however, in the presence of competitive inhibitor **4** (2.0 mM), under otherwise identical conditions, 82% of its catalytic activity is retained. These results indicate that a carboxylate group could be located in or very near the binding site of 43D4-3D12.

Affinity labels containing reactive epoxides react specifically with carboxylate groups of Asp or Glu and thiol groups of Cys residues. Affinity labels containing epoxides have been used to identify Glu and Asp residues responsible for catalysis in a number of enzymes, such as β-D-glucosidases, pepsin, lysozyme and phosphoglucose isomerase (Quaroni et al 1974). *Trans*-epoxide **6** was shown to inactivate 43D4-3D12 in a time-dependent manner (Fig. 3); in 30 hours, the specific activity of 43D4-3D12 was reduced to less than 5% of that of the native protein. After exhaustive dialysis following treatment with the affinity label no catalytic activity was recovered, indicating irreversible covalent inactivation. To distinguish between thiol-ether formation with a Cys residue and ester formation with Glu or Asp residues, we treated the inactivated antibody with 0.5 M hydroxylamine, pH 9.0. After exhaustive dialysis to remove any released affinity label, all activity was recovered, which strongly suggests that an ester linkage was responsible for attachment of the affinity label to the antibody. Radioactively labelled epoxide **6** has been synthesized and a peptide map of 43D4-3D12 is being prepared to identify the specific residue that reacts with **6**.

To determine the difference in mechanism between the antibody-catalysed reaction and the background reaction catalysed by acetate ion we did a kinetic isotope effect study. The deuterated substrate **7** was synthesized using

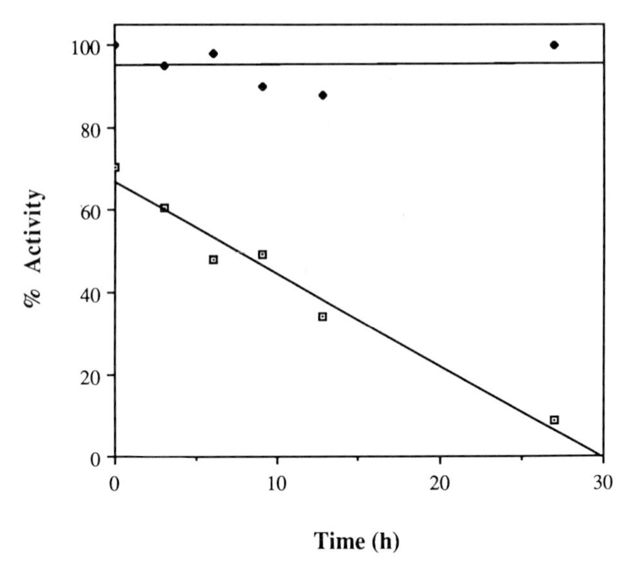

FIG. 3. The percentage activity versus time for inactivation of antibody 43D4-3D12 by **6**. (◆), control without **6**; (▫), 1 mM **6**. Activity is defined as $(v_{sample}-v_{background})/v_{unmodified}$ 43D4-3D12. The incubation buffer was 10 mM phosphate containing 150 mM NaCl pH 7.3. The stock assay buffer was 50 mM bis-Tris containing 100 mM NaCl pH 6.5. Assay conditions: to 340 µl of stock assay buffer, 150 µl of protein solution and 10 µl of 30.8 mM **2** (in CH_3CN) were added.

d_6-acetone. A Lineweaver-Burk plot of the elimination of **2** and **7** catalysed by 43D4-3D12 showed no difference in K_m between the deuterated and non-deuterated substrate (Fig. 4). The kinetic isotope effect on k_{cat} for this reaction was 2.35, a value indicative of a partially rate-limiting proton abstraction such as occurs in a classical E2 mechanism. Typical values of k_H/k_D for E2 mechanisms are 2–8, while the maximum for E1 mechanisms is 1.2 (for $E1cB_{ip}$) (Grossberg & Pressman 1960). For the acetate-catalysed elimination $k_H/k_D = 3.7$. This difference between the antibody-catalysed and uncatalysed reaction may result from the difference in pK_a between the catalytically active base in 43D4-3D12 ($pK_a = 6.2$) and that of acetic acid ($pK_a = 4.3$). A difference in the geometry of the transition state for the elimination reaction imposed by the combining site in the antibody-catalysed reaction could also account for the difference in isotope effects.

As mentioned above, hapten complementarity is not the only strategy for introducing catalytic groups into antibody combining sites. Both chemical derivatization and site-directed mutagenesis have been used to introduce a catalytic imidazole group or a thiol into antibody MOPC315 to generate effective esterolytic catalysts. The rate acceleration of coumarin ester hydrolysis achieved by the introduction of an imidazole via site-directed mutagenesis is

$(k_{cat}/K_m)/k_{imidazole} = 90\,000$ (Baldwin & Schultz 1989). The semisynthetically introduced thiol affords a rate acceleration of $(k_{cat}/K_m)/k_{thiol} = 60\,000$ (Pollack et al 1988). The rate acceleration achieved by 43D4-3D12 is in the same range, $(k_{cat}/K_m)/k_{acetate} = 92\,000$. By this analysis the different approaches appear to be equally successful in achieving high rate accelerations. This calculation, however, depends on the K_m for the substrate, which for ligands containing dinitrophenol is quite low in the case of MOPC315. The ratio of k_{cat}/k_{base} gives the effective molarity for the respective base in the combining site of the antibody. The effective molarities for the His mutant and thiol-containing derivatives of MOPC315 are 0.2 M and 0.073 M, respectively; the effective molarity for the β-elimination reaction catalysed by 43D4–3D12 is 16.8 M. This dramatic difference suggests that charge–charge hapten complementarity is a successful method for introducing catalytic side chains in close proximity to the reactive centres of bound substrates, leading to effective catalysts.

Covalent antigenicity

In our efforts to characterize the mechanism of the β-elimination reaction catalysed by antibody 43D4-3D12, the successful inactivation of the antibody by an epoxide-containing hapten mimic suggested another strategy for generating catalytic antibodies. If used as haptens, reactive affinity labels could select for combining sites on the basis of the nucleophilicity of binding site residues. We have termed this approach covalent antigenicity, because it depends on the covalent nature of the interaction between the hapten and the receptors on B cells in the immune system. This strategy could lead to antibodies that contain a variety of catalytic groups in the active site, including nucleophiles (Ser, Cys, His), electrophiles (Lys, Arg), acids (Glu, Asp) and bases (Arg, Lys, His). Enzymes which use this strategy include serine proteases, β-glucosidases and *cis/trans* isomerases.

Affinity labels are powerful tools for mapping protein binding sites, introducing spectroscopic probes and delivering reactive groups into protein binding sites. A great deal is known about the reactivity of various labelling groups towards amino acid side chains in proteins. For instance, epoxide groups react specifically with deprotonated carboxylate side chains or with thiol groups of Cys residues (McCaul & Byers 1976); maleimides or disulphides react solely with cysteines; α-bromo ketones react with amines, thiols and carboxylates; and diazoniums react with Tyr (Fersht 1985). Just as antibodies elicited to positively charged ammonium groups contain negatively charged carboxylate groups, antibodies raised against epoxides should contain reactive carboxylate groups or thiols in their combining sites. The stability of the affinity label should be such that it is not degraded *in vivo* so quickly that the immune system responds to the hydrolysed form of the hapten and not the reactive form.

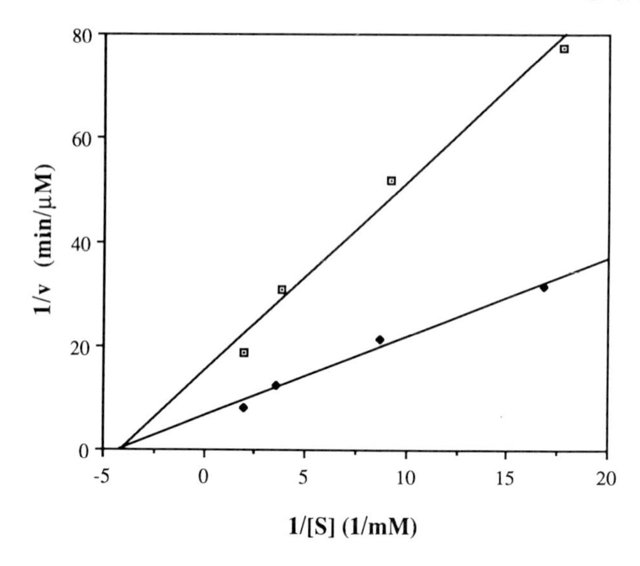

FIG. 4. Lineweaver–Burk plot of elimination of **2** (◆) and **7** (□) catalysed by antibody 43D4-3D12. Assay conditions were as for Fig. 1.

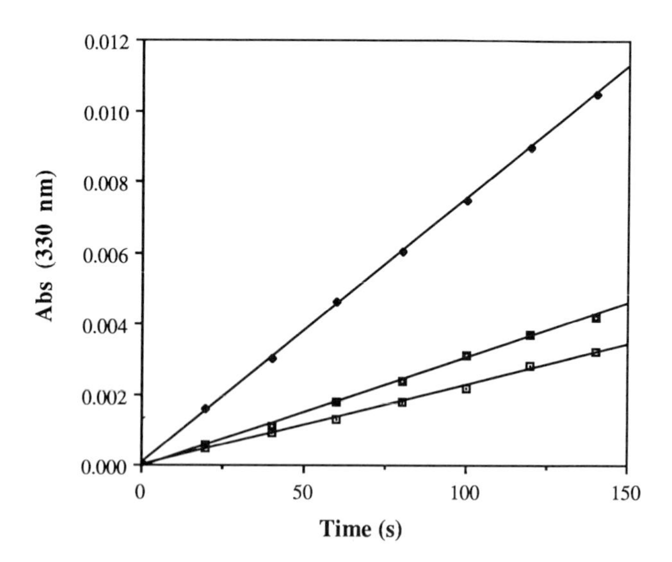

FIG. 5. A_{330} nm versus time plot for β-elimination of **2** to **3** catalysed by antibody 8M194.2. (◆), 8M194.2 at 2 μM; (■), 2 μM 8M194.2 containing 100 μM **8**; (□), background reaction. Assay conditions were as for Fig. 1 but pH 6.5. Substrate stock was 30.8 mM; final reaction concentration was 616 μM.

As a first test of the strategy of covalent antigenicity, antibodies were elicited against epoxide **8**. A carboxylate was incorporated into affinity label **6** used to label antibody 43D4-3D12 to facilitate attachment to the carrier proteins, bovine serum albumin and keyhole limpet haemocyanin. The stability of hapten **8** was measured by proton NMR in D_2O, the half-life for decomposition was 43 days. Our standard immunization protocol is 23 days, thus no significant decomposition of the hapten should occur before a high titre response can be elicited. Six monoclonal antibodies specific for hapten **8** were isolated. Two of these have shown preliminary indications of catalysis of the β-elimination reaction **2** to **3**. The time courses of the reaction in the presence of antibody 8M194.2 and the background reaction are shown in Fig. 5. The catalysed reaction is inhibited by both hapten **8** and the ammonium-containing hapten **1** that was used to elicit 43D4-3D12. These results strongly suggest that the catalysis occurs in the combining site of the epoxide-elicited antibody.

To determine the importance of the epoxide portion of hapten **8** in eliciting the carboxylate, we synthesized two haptens as negative controls. Hapten **9** is an open-ring structural isomer of hapten **8** which contains all the same recognition elements but does not react with carboxylates or thiols. Of the six monoclonal antibodies isolated so far that are specific for **9**, none catalysed the desired β-elimination reaction. The second negative control hapten is alcohol **10**. This hapten was chosen because it contains the same *p*-nitrophenyl ring as the epoxide hapten and an alcohol at the position α to the ketone. Of the four monoclonal antibodies raised against alcohol **10** none showed any catalysis of the β-elimination reaction of interest. The fact that these closely related haptens do not elicit any catalytic antibodies further suggests that the crucial element in the structure of hapten **8** is the reactive epoxide, not the oxygen atom. These negative controls also show that an open ring form of the epoxide hapten has not elicited the catalytic antibodies.

Conclusion

The principle of electrostatic complementarity between antibodies and haptens has been used to generate catalytically essential amino acids in antibody combining sites. This strategy has led to antibodies capable of catalysing a β-elimination reaction, but is limited in that only charged amino acids can be elicited. To elicit uncharged residues with unique reactivities such as Ser, Cys, Lys and His, we have developed a new strategy termed covalent antigenicity. This approach relies on covalent bond formation between the hapten and surface receptors on B cells.

Preliminary results described here indicate this strategy is successful in producing catalysts of a β-elimination reaction. A detailed kinetic study of the reaction catalysed by one such antibody is in progress, including peptide mapping of the binding site, time-dependent inhibition by the hapten and the stereochemistry of the elimination reaction.

Acknowledgement

This work was supported by the Office of Naval Research grant number N00014-87-K-0256.

References

Baldwin E, Schultz PG 1989 Generation of a catalytic antibody by site-directed mutagenesis. Science (Wash DC) 245:1104–1107
Fersht A 1985 Enzyme structure and mechanism. W H Freeman, New York
Fersht AR, Renard M 1974 pH Dependence of chymotrypsin catalysis. Biochemistry 13:1416–1426
Grossberg AL, Pressman D 1960 Nature of the combining site of antibody against a hapten bearing a positive charge. J Am Chem Soc 82:5478–5482
Iverson BL, Cameron KE, Jahangiri GK, Pasternak DS 1990 Selective cleavage of trityl protecting groups catalyzed by an antibody. J Am Chem Soc 112:5320–5323
Jacobs JW, Schultz PG, Sugasawara R, Powell M 1988 Catalytic antibodies. J Am Chem Soc 109:2174–2176
Janda KD, Weinhouse M, Schloeder DM, Lerner RA, Benkovic SJ 1990 Bait and Switch strategy for obtaining catalytic antibodies with acyl-transfer capabilities. J Am Chem Soc 112:1274–1275
McCaul S, Byers LD 1976 Reaction of epoxides with yeast glyceraldehyde-3-phosphate dehydrogenase. Biochem Biophys Res Commun 72:1028–1034
Pollack SJ, Schultz PG 1987 Antibody catalysis by transition state stabilization. Cold Spring Harbor Symp Quant Biol 52:97–104
Pollack SJ, Nakayama GR, Schultz PG 1988 Introduction of nucleophiles and spectroscopic probes into antibodies. Science (Wash DC) 242:1038–1040
Pressman D, Grossberg A 1968 The structural basis of antibody specificity. Benjamin, New York
Pressman D, Siegel J 1953 The binding of simple substances to serum proteins and its effect on apparent antibody–hapten combination constants. J Am Chem Soc 75:686–693
Quaroni A, Gershon E, Semenza G 1974 Affinity labeling of the active sites in the sucrase-isomaltase complex from small intestine. J Biol Chem 249:6424–6433
Schultz PG 1989a Catalytic antibodies. Acc Chem Res 22:287–294
Schultz PG 1989b Catalytic antibodies. Angew Chem Int Ed Engl 28:1283–1295
Schultz PG, Lerner RA, Benkovic SJ 1990 Catalytic antibodies. Chem Eng News 68:26–40
Shokat KM, Schultz PG 1990 Catalytic antibodies. Annu Rev Immunol 8:335–363
Shokat KM, Leumann CJ, Sugasawara R, Schultz PG 1989 A new strategy for the generation of catalytic antibodies. Nature (Lond) 338:269–271
Spratt TE, Kaiser ET 1984 Catalytic versatility of angiotensin converting enzyme: catalysis of an α,β-elimination reaction. J Am Chem Soc 106:6440–6442
Tramontano A, Janda KD, Lerner RA 1986 Catalytic antibodies. Science (Wash DC) 234:1566–1570

DISCUSSION

Page: In the kinetic isotope effect studies, there was no incorporation of deuterium into *p*-nitrobenzyl acetone **4**. Was there any into the substrate **2**? If so, did it compete with elimination?

Shokat: I didn't test that. It would be difficult to measure incorporation into substrate **2**—it undergoes elimination to **3** very quickly.

Page: The fluoride was racemic; did you get elimination in only one enantiomer?

Shokat: It was hard to tell whether the elimination was stereospecific: to investigate that we need to make the fluoride stereospecifically.

Jencks: That elimination reaction would probably be E1cb with rate-determining deprotonation.

Shokat: Yes. The kinetic isotope effect of 2.35 is consistent with an $E1cb_{irreversible}$ mechanism, which has a range of kinetic isotope effect values of 2–8.

Jencks: Could you insert the corresponding compound with a chlorine leaving group into the active site of the antibody? That would test whether the reaction involves rate-limiting deprotonation.

Shokat: We tried to make that compound but it is not stable.

Janda: What was the pK_a of the two protons on the original substrate (β-fluoroketone)?

Shokat: We never measured that: I would assume it is within 1–2 pH units of that of acetone ($pK_a = 20$).

Janda: How much excess epoxide reagent over the protein did you use in the labelling?

Shokat: The protein concentration was about 5 μM; the epoxide concentration was 1 mM.

Janda: We've done some work with epoxides and they appear to alkylate the antibody indiscriminately.

Shokat: But the inactivation by epoxide can be inhibited by the hapten.

Janda: We have worked with diazoacetamide: if you remove the charge on the carboxylate, the antibody is precipitated.

Shokat: That's why we can't fully inactivate the antibody and have to be satisfied with about 20% residual activity. If we try to get complete inactivation, we precipitate the antibody.

Janda: What rate accelerations are you seeing for the β-elimination reaction **2** to **3**—comparing k_{cat} with k_{uncat}, not the second order rate constants?

Shokat: About 1500 at the optimal pH.

Janda: If you compare k_{cat} with the second order rate constant or k_{cat} with the reaction extrapolated to zero buffer concentration, you see greater rate accelerations. In our case (antibody 'bait and switch catalysis'), buffer is not involved in catalysis, so this comparison is valid. We see the same overall rate acceleration compared to solution of about 10^2–10^3.

Schwabacher: Did you correct for the difference in pK_a?

Shokat: No. We have some data for bicarbonate, which has a pK_a much closer to that of the carboxylate in the antibody combining site. That was hard to measure because of the loss of CO_2 in solution and other problems with the measurement of rate constants for bicarbonate-specific catalysis.

Benkovic: Was the curve describing the pK_a (Fig. 2) based on a theoretical curve or just drawn through the points?

Shokat: That was drawn through the points.

Benkovic: Then was the point at low pH a hint that there was another mechanism going on?

Shokat: We wanted to use the same buffer because there were some differences between the reactions in bis-Tris and in MES or other buffers. When we switched buffers to go below pH 5, we were worried that the antibody was being denatured.

Jencks: There are several reactions in which the expulsion of fluoride is catalysed by acid. That could account for such an intercept.

Leatherbarrow: Could you fit your data to an ionization curve and see where the intercept is at the limiting low pH value? If you assume a simple ionization, is it consistent with zero activity at very low pH?

Shokat: We could fit a simple ionization curve, but we haven't done that analysis yet.

Lerner: When you lower the pH, won't you lose binding also?

Shokat: The K_m was decreased at low pH, probably because of a reduction in charge repulsion.

Hansen: Do the antibodies you obtained by the second method, which were not catalytic, react with the epoxide?

Shokat: We haven't looked for covalent adducts with all of the antibodies. We know they bind BSA–hapten conjugates, but whether they have carboxylates or whether there is a cysteine, we don't know.

Benkovic: Was the epoxide a racemic mixture?

Shokat: Yes, it was.

Page: Is there any product inhibition? The α,β unsaturated ketone product is almost more electrophilic than the epoxide.

Shokat: No, there appears to be no product inhibition. The carboxylate must be right next to the α-proton and not in a position to do a Michael addition into the vinyl ketone.

Plückthun: In general, if you have a very reactive group in the hapten, wouldn't the fraction of antibodies elicited that actually bind to the antigen be lower because there will be many random hit events, and antibodies might be elicited that are completely unrelated to binding this particular antigen? If you have a reactive antigen that is covalently linked to the antibody you may activate a particular B cell just by hitting the antibody on the surface. Is this a significant event? Are you raising many antibodies totally at random, not related to binding this antigen?

Shokat: We are clearly not just eliciting antibodies at random. Of the six antibodies isolated so far that bind the antigen, two are catalytic.

Schultz: This is an affinity label, it is not a generic epoxide.

Plückthun: I know, but it might react on the surface. In general, thinking about reactive groups, could this be a problem? With a rather reactive label, could you increase the production of non-specific antibodies so much that you have a hard time screening?

Shokat: Yes, but we haven't had any problem identifying catalytic antibodies using this strategy. However, if we put diazoacetamide all over the carrier protein, keyhole limpet haemocyanin, we would get a mess. That is why we chose to use an epoxide.

Lerner: Peter, is this the epoxide with which you were going to try and induce an immune response via alkylation?

Schultz: Yes. There are many affinity labels for enzymes that have very high selectivity and have been used for peptide mapping. You don't want to use something that's so reactive it will not label a group in the binding site. You have to be chemically sophisticated in the choice of hapten, not from the point of a transition state analogue, but from the point of reactivity.

Janda: How do you know it is an epoxide ring-opening process, not just simple hydrogen bonding that could elicit this response? The epoxide itself could induce something that hydrogen bonds, such as a thiol. Could one distinguish those possibilities?

Schultz: We used the two ring-opened epoxides as controls, which is good as you can get.

Shokat: That is also why we chose a system that was well characterized. Ideally, it should be possible to pull out the same catalytic antibody by those two different methods, charge–charge complementarity and covalent antigenicity.

Schultz: If we get a cysteine, it will be pretty unequivocal.

Blackburn: Could your antibody catalyse the retro-aldol reaction?

Shokat: We fed 43D4-3D12 the aldol product of acetone with *p*-nitrobenzaldehyde for two reasons: to see if it would eliminate water and if it would catalyse a retro-aldol reaction, but it did neither. The antibody did bind the aldol product, since we saw inhibition of the **2** to **3** conversion at high concentrations. There is no reason to expect there to be an acid to protonate the alcohol, so it's not surprising that there was no elimination of water.

Blackburn: I am not referring exclusively to antibody 43D4-3D12 that catalyses the elimination reaction. Do any of the other five antibodies that you isolated catalyse this reaction? One would want the carboxyl group in a different position within the active site to catalyse a retro-aldol reaction.

Shokat: I tried all six antibodies for catalysis of the retro-aldol reaction.

Harris: For some drugs that have poor pharmokinetics, you can administer them as a prodrug, for example as an ester, which will be degraded to form the reactive species. Could you use this approach with a modified hapten?

Shokat: That would be like a sophisticated adjuvant.

Hilvert: You reported (Shokat et al 1989) that the affinity of the antibody for the *m*-nitro compound was only 10-fold less than that for the *p*-derivatives. Can you explain?

Shokat: k_{cat}/K_m was 1.4×10^2 M^{-1} min^{-1} for the *m*-nitro substrate and 1.4×10^3 M^{-1} min^{-1} for the *p*-nitro substrate. It does not show quite such good selectivity as does the transition state approach.

Hilvert: I was surprised by the relatively low selectivity because I would have expected the nitro aromatic moiety to be immunodominant and the antibody to discriminate between those substrates very effectively. That doesn't seem to be the case. Have you tried to titrate the active sites to eliminate the possibility that reactions are occurring non-specifically at multiple sites in the protein?

Shokat: We haven't tried to titrate directly, but it looks like there is complete inhibition with an equimolar amount of inhibitor. Also, the Dixon plot gives the correct concentration of active sites compared to antibody concentration.

Lerner: In the β-elimination study you had to give up some binding energy because there was a heteroatom in one situation and not in the other. So, to pick up some binding energy the nitro functionality may be important. The principle is interesting—if you are going to switch from substrates with homology, it may be useful to do as you did, and pick up binding energy elsewhere in the molecule.

Janda: I find it reassuring that you see the same type of numbers and chemical groups involved in different types of reactions as we do. But, I thought we would see better rate enhancements using catalytic antibodies than we are seeing. The next step should be to elicit two groups. With our substrates, e.g. phenol esters, the K_ms were about 1 mM. With a benzoate hapten we believe we have elicited an arginine in the antibody binding pocket. With the same substrate, the K_m is a little lower, around 250 mM, similar to those you find.

Presumably, it is possible to put two groups into the binding site of the antibody. Some of the earlier literature says that if you put two groups too closely together, you won't get recognition of both charges, which could be another problem (Freedman et al 1968 and reference therein). It worries me that the antibody then won't bind substrate at all: you can't rely on nitrophenyl groups as recognition elements for all antibody-catalysed reactions.

Blackburn: On this two-carboxylate problem, could you separately induce one carboxylate on the light chain and one on the heavy chain, then create the antibody you want using Bill Huse's recombination system?

Janda: You can put them in but it would be difficult to find a substrate that way.

Schultz: We are attempting to generate two charges in a number of antibody systems, including proteases and glycosidase haptens. There are other combinations: for example, strain plus charge or entropy reduction plus charge, which may be easier to achieve.

Another way to generate a second carboxylate group in an antibody binding site is to use a good screen or selection procedure.

Lerner: Do we have to copy aspartate proteases? Are there more interesting ways to achieve catalysis?

Schultz: The rate accelerations achieved by current catalytic antibodies are in the order of 10^3–10^6. If one dissects an enzyme, such as the Staphylococcal nuclease, one realizes that Nature uses a combination of factors to achieve high reaction rates.

Janda: When making these comparisons between enzymes and catalytic antibodies, the charge complementarity idea or what we call 'bait and switch catalysis' may not be useful by itself, you may need to use this approach in conjunction with transition state stabilization etc. At this point, if you used both charge complementarity and transition state stabilization in the antibody procurement scheme and obtained large rate accelerations, you would not know if it was due to one or the other or both.

Hilvert: Doesn't this all have to be transition state stabilization? That's the only way you can get catalysis, by definition. You may change the mechanism by introducing other catalytic groups but it's still transition state stabilization. That's why carrying out the uncatalysed reaction with acetate is important.

Janda: From the earlier attempts to raise catalytic antibodies, where we were using tetrahedral haptens (i.e. phosphonates) one could say that. More recently, we have used a 'bait and switch' approach and in this instance the scissile bond in the substrate is replaced by a neutral isostere; hence I think it's a little different.

Jencks: Transition state stabilization is slightly ambiguous because it is never quite clear whether one should include substrate stabilization. Sometimes the term is applied to the difference between the bound substrate and the transition state; other times it's applied to k_{cat}/K_m, which represents the difference between the substrate in solution and the transition state.

Hansen: It is fine to try to deconvolute geometric stabilization from electrostatic stabilization. Kim, are you saying that it may be impossible, for example, to make a catalytic antibody triose phosphate isomerase, which has a tiny substrate but possibly three charges in the active site?

Janda: I don't want to say it can't be done. But I think you will need a larger haptenic molecule with more recognition sites available. I believe you would not lose an enormous amount of binding energy in cases where you used an uncharged substrate and the antibody was elicited from a multicharged hapten.

Schultz: The other issue is that by juxtaposing the charges one carboxylate may be deprotonated and the other protonated.

Huse: One can make organic molecules to try to select a particular antibody, but if the immune system doesn't generate enough antibodies against that antigen to include one with catalytic activity, one cannot select it.

Schultz: We can use random mutagenesis to increase the diversity.

Huse: But can that be done in a step-wise fashion?

References

Freedman MH, Grossberg AL, Pressman D 1968 Evidence for ammonium and guanidinium groups in the combining sites of anti-*p*-azobenzenearsonate antibodies—separation of two different populations of antibody molecules. J Biol Chem 243:6186–6195

Shokat KM, Leumann CJ, Sugasawara R, Schultz PG 1989 A new strategy for the generation of catalytic antibodies. Nature (Lond) 338:269–271

General discussion I

Paul: A substrate molecule may provoke catalytic antibody formation. Have control experiments been done using the ground state as the immunogen and screening for catalytic antibodies by binding of a transition state analogue?

Janda: We have preliminary results where we immunized with an α-methyl benzyl ester. That compound is stable over the period needed to generate an immune response. Early in the response we obtained antibodies that bound the ground state and the transition state analogue equally well. Later, we found only antibodies that bound the ground state. We were looking for antibodies that bound the transition state more tightly even though they were raised against the ground state. The problem was that most of the antibodies we obtained were IgM. These are notoriously difficult to purify, so we are not certain that we have catalytic antibodies at this stage.

We have also used the charge complementarity approach. We used two different uncharged haptenic molecules as controls. Each possessed a hydroxyethylene moiety in place of the scissile bond to be broken. We found no catalytic antibodies; however, Peter Schultz has found that a *p*-nitrohydroxyethylene hapten did induce catalytic antibodies.

Benkovic: Even though we have molecular orbital calculations and molecular graphics, the study of catalytic antibodies is still very empirical. We should use the diversity of the antibody system: there are some very powerful lessons from protein engineering that point this out. There have been many attempts to take a protein and change its substrate specificity by protein engineering. With antibodies we have the advantage of starting with the desired specificity. You can change the specificity of an enzyme, but, at present, if you change dihydrofolate reductase to lysozyme, you end up with lysozyme. You usually make amino acid substitutions that Nature has already made. The diversity of the immune system provides an opportunity to see if there are other chemical solutions.

In enzymology there are several general solutions to the problem of amide bond cleavage; but not many, considering the possible diversity of the system. We might be able to find some new catalytic pathways that we would never be able to predict. Mutagenesis of the antibody binding site may, by analogy, help to fine tune catalysis, but the desired properties of substrate specificity and catalytic chemical transformations reside in cleverly induced antibody libraries.

Burton: Where are we going to get these diverse libraries? What's just beginning to emerge from gene sequencing is that if you immunize, you lose a lot of the diversity. You probably do want to immunize with the transition state analogue to get the system going. But to tap the repertoire, we need to introduce probably several rounds of mutagenesis.

Benkovic: Would you do random mutagenesis?

Burton: I think so.

Plückthun: The great attraction of the antibody system is the diversity, and potentially its generality. However, the amount of work we can do is not infinite. Therefore we need a better idea of what we are aiming for: more of the successful catalytic antibodies that we have discussed should be looked at structurally. If we just treat catalytic antibodies phenomenologically, do we learn anything to make our next approach more efficient? Protein engineering, especially structural work, is very important.

Jencks: We need some quantitative evaluation of possible contributions of particular catalytic mechanisms.

Benkovic: We also need an extensive study of some of these catalysts. You can't just compare k_{cat}/K_m to the spontaneous rate—you don't know if you are comparing oranges and apples. We don't know if we are looking at chemistry, product desorption or conformational changes in between, all of which could be rate limiting in substrate turnover by the antibody catalyst.

Huse: Immunization with a transition state biases the system towards a particular mechanism which includes that transition state. If you could screen initially for the catalysis rather than for binding to the transition state, then that problem goes away.

Schultz: There are pretty sensitive assays for screening for catalysis directly on agarose plates. We are looking at fluorescence energy transfer; chemiluminescence is another option, pH indicator plates are a third. If you are going to screen a library directly, you have to make sure that the reaction is not catalysed by a bacterial enzyme.

Hilvert: The nitroanilide-cleaving antibody mentioned by Kim Janda was identified by screening a large number of clones in an ELISA plate with a colorimetric assay.

Paul: The answer might be selective trapping of antibodies onto a solid support followed by screening for catalysis. That would avoid background activity due to enzymes and so forth.

Benkovic: In the near future there will be genetic methods for screening. There will be developments with fluorescence methods that will be very sensitive. There will also be increased expression of the antibodies.

Green: The use of monoclonal antibodies as catalysts requires an efficient screening procedure to select those hybridomas that secrete catalytic antibodies from the much larger number of clones secreting antibodies which only bind the hapten used for immunization. We have developed a simple, general procedure using what we have called 'short transition state analogues' (Tawfik et al 1990). Such substances are truncated analogues of the immunizing hapten

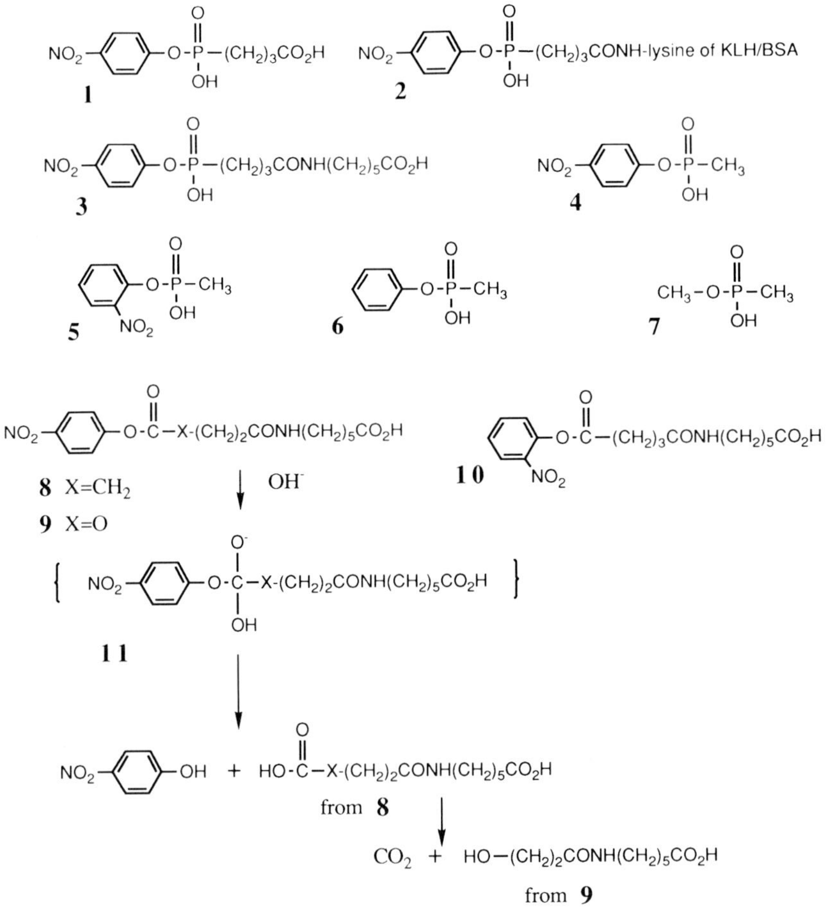

FIG. 1. (*Green*) Chemical structures of hapten **1**, conjugate **2**, substrates **8–10**, and inhibitors **3–7** used, and of the ester **8** and carbonate **9** hydrolysed by the antibodies. KLH, keyhole limpet haemocyanin; BSA, bovine serum albumin.

(the transition state analogue for the reaction to be catalysed by the antibody) which, ideally, contain all the unique elements of the transition state and should bind strongly to the catalytic monoclonal antibodies.

The structure of the short transition state analogue should contain the minimum number of elements common to substrate and transition state (on the basis of the definition of catalysis: strong binding to the transition state; weak binding to the substrate). It should also contain few elements common to transition state and product(s) in order to ensure that the catalytic monoclonal antibodies display turnover (minimization of product inhibition). The assay is rapidly and easily performed using standard ELISA techniques for monoclonal antibody binding; we measure the competitive inhibition of binding by the short transition state analogue and the hapten.

Monoclonal antibodies raised against the hapten **1**, linked to a protein carrier to give immunogen **2** (Fig. 1), were expected to catalyse the hydrolysis of the analogous ester **8** and carbonate **9**. The transition state (or transient state) for this reaction may be depicted as **11**, which is mimicked by the transition state analogue **1**, the hapten used to elicit monoclonal antibodies, or hapten **3** where 6-aminocaproate replaces the lysine residue of the protein carrier. We reduce the transition state analogue to the smallest structure that will still bind to the antibodies produced and, of course, retains the phosphonate function. In this case, we chose *p*-nitrophenyl methylphosphonate **4**.

After immunization with the KLH–hapten conjugate **2** and fusion, the supernatants of the resulting hybridomas were screened for binding to the BSA–hapten conjugate **2**. Fig. 2 shows the method used to select the monoclonal antibodies. All the antibodies that were bound by the hapten **3** were considered for further study. We examined those antibodies within this class that had affinity for the 'short' transition state analogue **4**. To ensure that we would be able to get high-purity antibodies (the presence of trace quantities of enzymes is disastrous if one is seeking antibodies that catalyse a reaction which may also be catalysed by enzymes present in the media), at this stage we selected only antibodies that bound to protein A. 14 such antibodies with affinity for the short transition state analogue were grown in large quantity as ascites then purified by protein-A affinity chromatography. Each antibody was tested for catalytic activity with respect to the ester substrate **8**, in the presence and absence of the hapten **3**. Reactions were compared to those of a non-relevant antibody control. The 14 antibodies can be assigned to three groups with respect to affinity for the short transition state analogue: those having strongest, strong, and medium affinity for **4**. Of the five clones in the first class (90–100% inhibition of conjugate binding by **4**), three catalysed the hydrolysis of **8**. Of the two clones displaying 80–90% inhibition of binding, one was catalytic; the seven clones showing 60–80% inhibition of binding were not catalytic.

We examined short transition state analogues which were smaller than **4**, such as phenyl methylphosphonic acid **6** and methyl methylphosphonic acid **7**, but these did not exhibit sufficient binding to the antibodies to be useful for indicating catalytic potential. The affinity of the antibody for the solid phase hapten–protein conjugate is very high, even relative to the soluble hapten, and the design of an effective short analogue must take this into account. It has to be experimentally verified that the short transition state analogue used has sufficient affinity for the antibody to inhibit binding and thus be detectable using the competitive inhibition enzyme assay.

Table 1 displays the catalytic constants for the reactions with the ester **8** and carbonate **9**. We find it interesting that the antibodies, all of which hydrolyse the ester, show such varied activity with the carbonate. With monoclonal

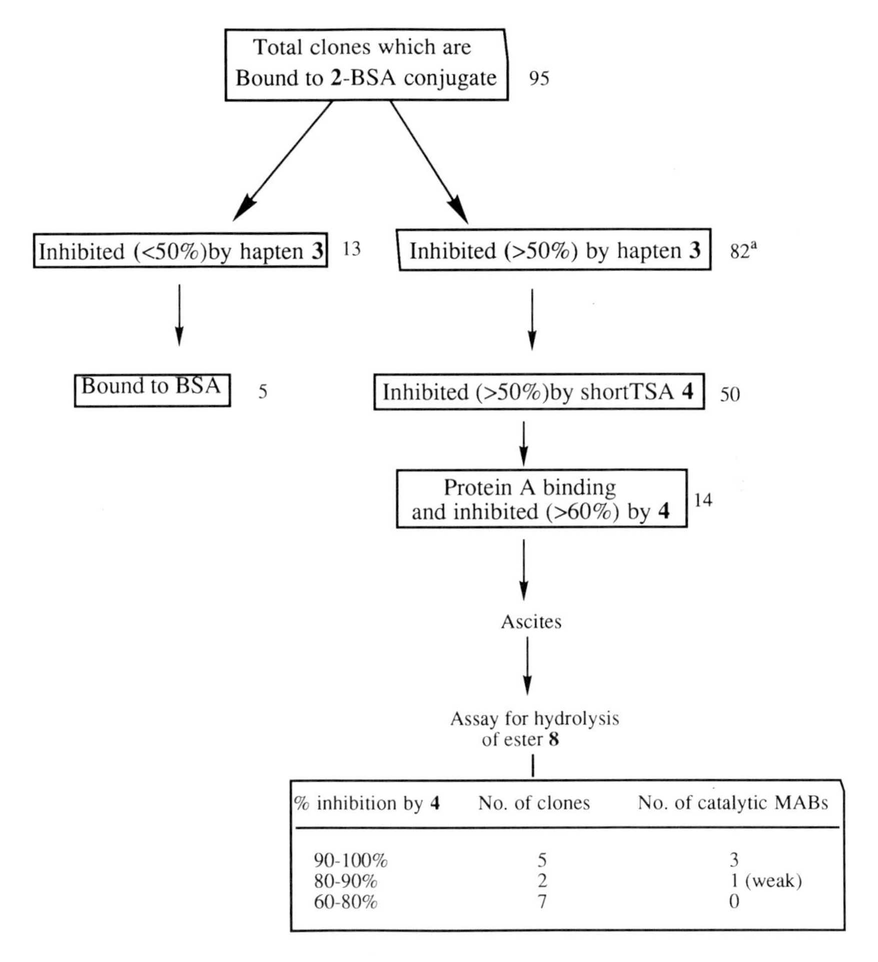

% inhibition by **4**	No. of clones	No. of catalytic MABs
90-100%	5	3
80-90%	2	1 (weak)
60-80%	7	0

FIG. 2. (*Green*) Scheme for the selection of catalytic monoclonal antibodies. [a]Had all these 82 clones been screened for protein A binding and propagated as ascites, a total of 30–40 would have been obtained as purified monoclonal antibodies to be assayed for catalytic activity. BSA, bovine serum albumin; TSA, transition state analogue.

antibody CNJ 157 the rate enhancement is appreciably greater than for the ester; CNJ 123 does not catalyse the carbonate reaction. With CNJ 19 the carbonate hydrolysis is a stoichiometric reaction. We believe that the drop-off in reaction both with the carbonate **9** and ester **8**, as reflected by the number of turnovers shown in Table 1, is due to a covalent reaction of the substrates with amino acid residues at or near the combining site. This is not surprising since *p*-nitrophenyl esters are active esters, known to acylate amino or phenolic functions. This problem of covalent inhibition may be solved using a substrate such as the *p*-iodo analogue of **8** for which K_m should not be significantly

TABLE 1 (*Green*) Kinetic parameters of the catalytic monoclonal antibodies

			Ester 8			*Carbonate 9*	
Monoclonal antibody	K_m[a] *mM*	*Ki*[b] *μM*	k_{cat} min^{-1}	*No. of turnovers*	k_{cat}/k_{uncat}	*No. of turnovers*	k_{cat}/k_{uncat}
CNJ 157	0.11	3.4	2.39	50	0.97×10^4	6	1.60×10^4
CNJ 123	0.08	57.0	0.79	30	0.32×10^4	—	—
CNJ 2	0.06	ND	1.1	38	0.46×10^4	4	0.80×10^4
CNJ 19	*ca.*0.30[c]	0.09	0.63	28	0.26×10^4	2	0.63×10^4

[a]The K_m values were determined using **8** as substrate.
[b]The K_i values were determined using **4** as inhibitor.
[c]The K_m for CNJ 19 was determined at substrate concentrations up to 1 mM; attempts to increase the substrate concentration to the K_m range resulted in inactivation of the antibody. The value given is therefore only an estimate.

different but which should be a much less effective acylation reagent. The hydrolysis reactions are specific for the *p*-nitrophenyl substrates, as expected from the way they were selected. The reaction with the *o*-nitrophenyl ester substrate **10** is not catalysed.

The problem of establishing that the observed reaction is due to antibody catalysis and not to some extraneous enzyme or other impurity has been addressed several times at this symposium. In addition to specific inhibition of the reaction by the hapten, and substrate specificity of the reaction, we have correlated the specificity of antibody binding to a panel of inhibitors (**3**, **4**, **5** and **6**) with the independently measured specificity of inhibition of catalysis. As seen in Fig. 3, the relative affinities of these inhibitors for the catalytic antibody are exactly as expected for antibodies raised against structure **2**: they are identical for inhibition of catalytic activity. Demonstration of such a correlation minimizes the probability of an enzyme impurity having the same pattern of binding and catalysis inhibition; one could use additional inhibitors to lower the likelihood of an enzyme impurity even further.

Eventually, we all hope that it will be possible to screen hybridoma supernatant solutions directly for catalysis. Since this is not yet possible, the short transition state analogue approach should be a useful aid for obtaining catalysts which are effective and display turnover. This is part of a general methodology which advocates the wider application of ELISA binding using substances designed to select for desired qualities before expanding clones and ascites production. This approach may also be useful in the selection of promising

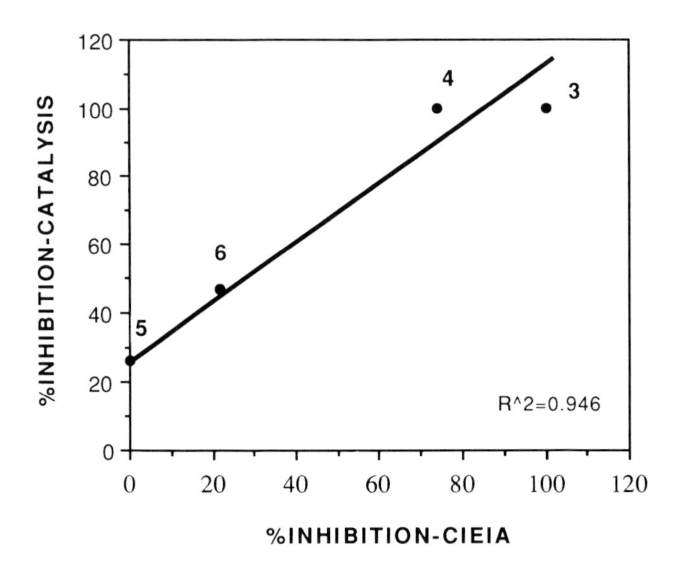

FIG. 3. (*Green*) Correlation of binding of inhibitors **3–6** to a catalytic antibody and their ability to block its catalytic activity.

antibodies from the new libraries being developed, such as combinatorial libraries.

Benkovic: When using the *p*-nitro antigens, the problem of severe product inhibition arises very early. So if you are going to use Michaelis–Menten-type equations, you will have to look at a very small percentage of the reaction. The results in my paper (this volume) were all obtained using stopped-flow techniques for the *p*-nitro system because that was the only way we could get a small enough turnover to avoid product inhibition. In the case where you saw no turnover with the carbonate **9**, perhaps there was a single turnover and the product was tightly bound to the antibody.

Green: The products of the ester hydrolysis reaction, *p*-nitrophenol and the carboxylic acid, were tested as inhibitors of binding and of hydrolysis. None of the antibodies was inhibited by the carboxylic acid product; one was weakly inhibited by the *p*-nitrophenol. Thus, the lack of turnover using the carbonate substrate with monoclonal antibody CNJ 19 was not due to product inhibition. The absence of turnover results from acylation of an amino acid residue in or near the antibody combining site.

Lerner: I agree in principle that this is a powerful approach. Let me divide the antigens into parts that are mechanistically interesting and those that are not. The phosphonate is the mechanistically interesting part; the nitro functionality is the uninteresting part. If you remove the nitro group when you screen and lose all activity in the ELISA, is your short transition state analogue short enough? In other words you are screening for the uninteresting part. You should remove the nitro group and the benzene ring and screen just for the phosphonate or phosphonamidate.

Hansen: It is interesting that Dr Green is removing the floppy side of the molecule—the linker side, the methylene chain. The nitro side is not very floppy and there may be more of a lever-type effect. The flexible part of the molecule may be able to adopt several orientations which are 'uninteresting', but the relatively rigid part of the molecule may have to be in the correct orientation.

Green: I believe we could also have used a short transition state analogue where the nitro group or the nitrophenyl group was absent—that is, include only the glutaramide side—and still obtained catalytic antibodies. However, since the short transition state analogue has to have sufficient binding affinity for the antibody and the intrinsic binding of the glutaramide is probably much smaller than that of the *p*-nitrophenyl group, one might have to include a portion of the immunogen to ensure sufficient binding.

Janda: It seems to me that you are just cutting down the size of the linker. I don't see the overall generality of this. Typically, when using multisubstrate-type analogues, we screen both parts and look for binding that way.

Green: Use of the short transition analogues is simply a way of rapidly identifying those antibodies that bind tightly to the transition state and not to the substrate(s) or product(s). We want the shortest molecule that retains the essential structural elements of the transition state and has sufficient affinity for the antibody.

Janda: But you are still screening for binding to the transition state analogue.

Green: We initially select those clones whose antibodies bind the hapten—the entire transition state analogue. We then minimize the number of clones that has to be examined further by choosing only those that show the tightest binding to the short transition state analogue.

Schultz: We have done a slightly different experiment. We generated antibodies to a nitrophenyl phosphonate hapten and asked what would happen if we generated antibodies to a phenylphosphonate hapten. Mark Martin will describe these antibodies in his paper (this volume).

Hansen: If you screen for just tremendously good binding of the antibody to, for example, the phosphate, I don't see how that's different from immunizing with phosphate. If we look at enzymes as a model, we want to achieve the best possible binding of the whole molecule in that precise geometry. Your data are compelling, but I am not sure that one wants to ignore the ends of the transition state analogue.

Janda: A chemist might ask: why include the extra spinach in the molecule unless it is required in the first place? Why not just design a simple molecule?

Schwabacher: Because of product inhibition. You want the whole molecule to be recognized and you want the whole molecule to be recognized more easily than is just a small piece of it. You want the major portion of the binding energy to be available for a certain substructure. The other purpose is as a place-holder. If you want a catalyst that requires a substructure but permits substituents in a certain region, immunization with the larger molecule ensures that the binding site has room for substituents; screening with the smaller one selects those that don't require them and are likely to allow variation of substituents.

Hansen: If you could raise an antibody to phosphate, do you think it would be catalytic? I would predict that it would not be very effective as an esterase because there is no binding energy apart from the 'interesting piece' to drive the reaction.

Lerner: We should not overlook the potential for the study of catalytic antibodies to give something back to physical organic chemistry. For example, consider the question of whether ferrochelatase reorientates the nitrogens of the porphyrin: the catalytic antibody field was able to show that taking those nitrogen atoms out of plane is important for the catalytic activity.

Hilvert: We must, however, be careful about inferring too much from our ability to produce catalytic antibodies. We may design an antigen according to a particular mechanistic bias and that antigen may elicit a catalytic antibody, but it may have produced it for reasons that we don't understand and that have nothing to do with our design.

References

Benkovic SJ, Adams J, Janda KD, Lerner RA 1991 A catalytic antibody uses a multistep kinetic sequence. In: Catalytic antibodies. Wiley, Chichester (Ciba Found Symp 159) p 4–12

Martin MT, Schantz AR, Schultz PG, Rees AR 1991 Characterization of the mechanism of action of a catalytic antibody. In: Catalytic antibodies. Wiley, Chichester (Ciba Found Symp 159) p 188–200

Tawfik DS, Zemel RR, Arad-Yellin R, Green BS, Eshhar Z 1990 A simple method of selecting catalytic monoclonal antibodies that exhibit turnover and specificity. Biochemistry 29:9916–9921

Screening combinatorial antibody libraries for catalytic acyl transfer reactions

Lakshmi Sastry, Monica Mubaraki, Kim D. Janda, Steve J. Benkovic and Richard A. Lerner

Department of Molecular Biology & Chemistry, Research Institute of Scripps Clinic, La Jolla, CA 92037, USA

Abstract. A bacteriophage λ vector system for the expression of Fab fragments from the mouse antibody repertoire in *Escherichia coli* has been described. We have used this system to generate a catalytic antibody from a combinatorial antibody library. Monoclonal antibody 43C9 was raised against a transition state analogue of the hydrolysis of carboxyamide. mRNA from hybridoma cells expressing this antibody was cloned into phage λ and clones that expressed the mRNA for either the heavy or the light chain of the antibody were isolated. These individual libraries were then crossed to generate a combinatorial library in which clones coexpressed the heavy and light chains. This library was screened for antibodies/Fab fragments that bound to the original antigen with high affinity. DNA sequencing showed that these fragments were the same as those in antibody 43C9. Three different clones were found to catalyse the hydrolysis of carboxyamide. More efficient expression vectors and improved screening techniques should lead to the isolation of many more catalytic antibodies from combinatorial antibody libraries.

1991 Catalytic antibodies. Wiley, Chichester (Ciba Foundation Symposium 159) p 145–155

Monoclonal antibodies are used extensively in various fields of biology and medicine. Some important applications include the investigation of cellular mechanisms, the isolation of interferons, cancer research, clinical diagnosis and gene product analysis. The generation of monoclonal antibodies with specific catalytic functions is an emerging technology that combines the high specificities of antibodies with chemical catalysis. A number of reactions have been successfully catalysed by monoclonal antibodies (for review see Lerner & Benkovic 1988). The production of homogeneous antibodies for catalysis is entirely dependent on the hybridoma technology; but this is an inefficient method

Heavy chain vector - λHc2

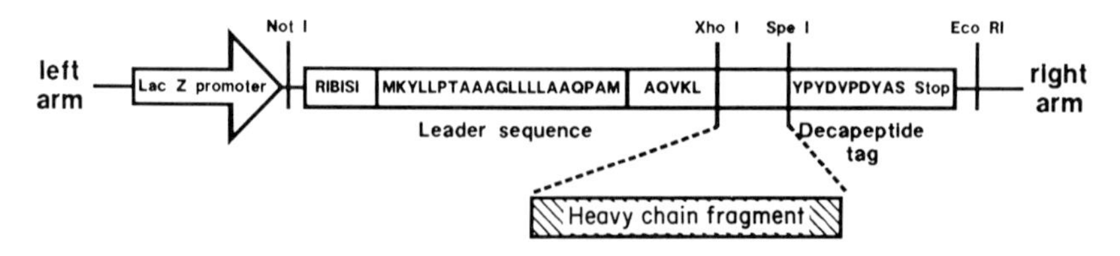

Light chain vector - λLc1

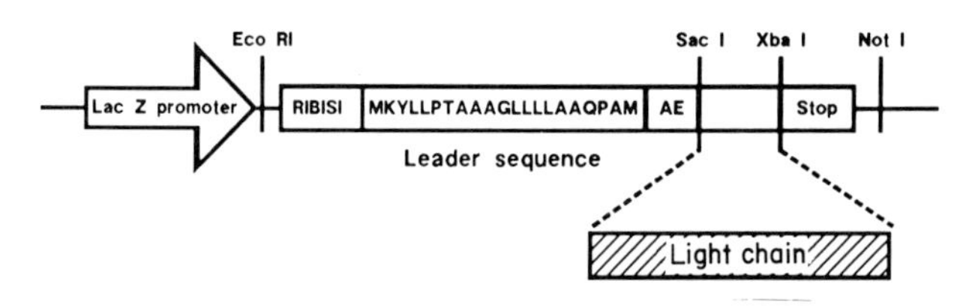

Combinatorial construct

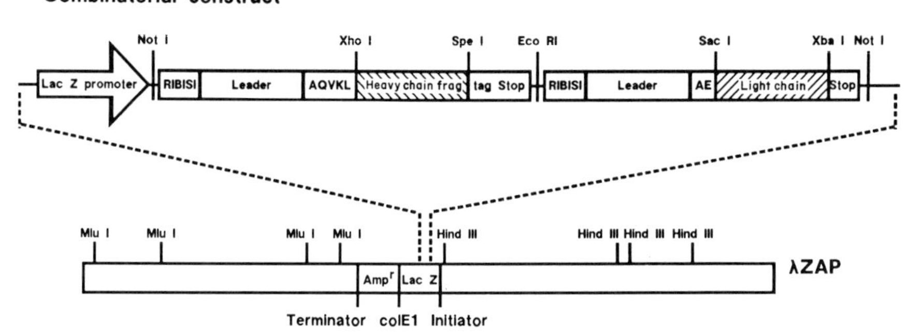

FIG. 1. Combinatorial bacteriophage λ vector system for expression of Fab antibody fragments. The LC1 vector is for cloning PCR products of mRNAs that code for κ light chains; the HC2 vector is for cloning PCR products of mRNAs coding for heavy chain Fd sequences. The combinatorial constructs that can express Fab fragments are generated by cutting light and heavy chain DNA at the antisymmetric *Eco*R1 site of each vector, followed by religation of the resulting arms.

for surveying the immunological repertoire and limits the number of catalysts that can be obtained. We have developed a system using bacteriophage λ to clone and express a combinatorial library of Fab fragments of the mouse antibody repertoire in *Escherichia coli* (Fig. 1) (Sastry et al 1989, Huse et al 1989). The system allows rapid and easy identification of monoclonal Fab fragments in a form suitable for genetic manipulation. However, it remains to be shown that such combinatorial libraries can be used to produce catalytic Fab fragments. In this paper we demonstrate the generation of a catalytic antibody from a combinatorial antibody library.

Using the λ phage system we generated an Fab combinatorial library from the spleen of a mouse immunized with phosphonamidate **1** (NPN), a transition state analogue for the hydrolysis of carboxyamide substrate **2** (Fig. 2). Screening the library with the antigen, NPN, linked to bovine serum albumin (NPN-BSA) resulted in the identification of a number of Fab fragments that bound to the antigen in a competitive manner. To find efficient catalysts for the hydrolysis of the nitroanilide **2**, we screened the Fab combinatorial library directly for catalysis. The induced phage libraries were incubated with nitrocellulose filters saturated with the substrate, or the substrate was added directly to agar containing cells infected with the phage before they were poured onto a plate. Unfortunately, these approaches were unsuccessful because of the chemical nature of the reaction as well as the limited amount of Fab that is secreted by the phage molecules. It has previously been observed that catalysis of hydrolysis of the amide **2** occurs at 37 °C with high concentrations of an antibody (Janda et al 1988).

The high concentrations of antibody required for catalysis are difficult to achieve directly on the phage surface. Also, the product of the hydrolysis, *p*-nitroaniline, is diffusible and is hard to observe either directly on phage

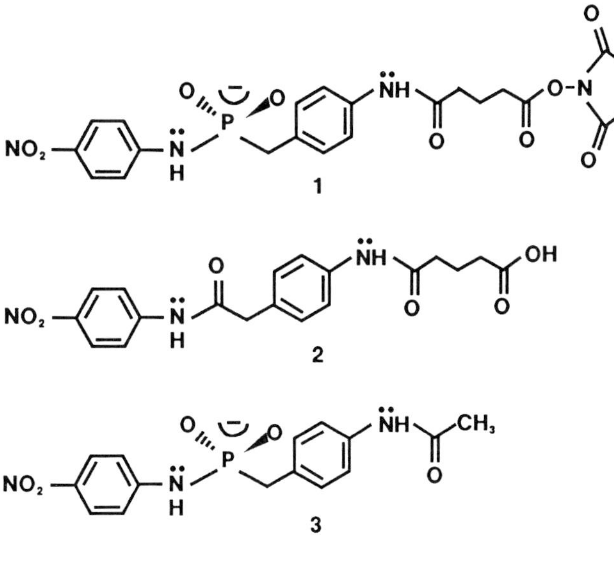

FIG. 2. The transition state analogue phosphonamidate **1** (NPN) which was used to induce antibodies that hydrolyse the carboxyamide substrate **2**. The phosphonamidate functionality mimics the stereoelectronic features of the transition state for hydrolysis of the amide bond. The transition state analogue **3** is an inhibitor of the reaction.

plaques or on nitrocellulose filters. Because of these practical limitations, we decided to screen initially for Fab fragments that bound to NPN and then for those that showed catalytic activity. As an essential first step, we cloned and expressed a monoclonal antibody (43C9) that catalyses the amide hydrolysis in the phage system (Janda et al 1988). Besides being an internal control, the expression of the monoclonal antibody in phage also allows the study of its structure and mechanism of catalysis. Mutagenesis and chain-exchange experiments can be easily performed on the cloned antibody to improve its catalytic activity.

Methods

Total RNA from 10^7 43C9 hybridoma cells was isolated as described (Chomczynski & Sacchi 1987). The mRNAs were purified on an oligo dT column, then amplified using the polymerase chain reaction to obtain separate pools of heavy and light chain DNA (Sastry et al 1989, Huse et al 1989). Amplification of heavy chain DNA was performed with eight different 5′ primers and a 3′ primer specific for the IgG2b isotype. Light chain DNA was similarly amplified with five 5′ primers and a κ chain-specific 3′ primer. Heavy and light chain libraries were generated in phage λ and crossed to obtain an Fab combinatorial library (Huse et al 1989). This library was then screened with NPN-BSA labelled with ^{125}I and Fab fragments that bound the antigen were identified (Huse et al 1989). These Fab fragments were excised using helper phage (M12 mp8) and McBlue cells and plated on LB/ampicillin plates (Short et al 1988). Colonies on the plates represented the excised plasmid carrying the cloned heavy and light chain pieces.

Individual clones were grown up and their protein products isolated using an affinity column made from anti-$(\text{Fab}')_2$ coupled to Sepharose beads. Purified Fab was dialysed for 4–6 hours against ATE (Aces, Tris, ethanolamine) buffer, pH 9.0, concentrated to 1–3 μM solution, and used for catalysis. Catalysis was performed at 37 °C in ATE buffer at pH 9.0 with the 1–3 μM Fab solution and a saturating amount (1 mM) of substrate **2**. Sequencing of the positive clones was as described by Sanger et al (1977).

Results

PCR amplification of heavy and light chain DNA resulted in bands of about 700 bp as analysed by agarose gel electrophoresis. A number of different primers were used for amplification from the hybridoma cells, because these may contain other non-functional heavy or light chains and restricted amplification may result in the cloning and expression of the wrong chains. To avoid this problem, we pooled the amplified DNA from the heavy and the light chains, then cloned each pooled fraction into the expression vector. Cloning of heavy chains resulted in 2×10^6 recombinants; the light chain library contained 5×10^5 recombinants.

Screening of the heavy chain recombinants with an antibody raised against a conserved 10 amino acid sequence in the heavy chain showed that 90% of these were expressing the decapeptide and therefore the heavy chain. Anti-κ antibody screening of the light chain library indicated that 60% of the clones were expressing κ light chains. The combinatorial library consisting of 2×10^7 recombinants was screened with the anti-decapeptide and anti-κ antibodies; 65% of the clones coexpressed heavy and light chains.

The Fab library (3000 plaques/plate) was then screened with iodinated NPN-BSA and positive clones were identified after a three-day exposure. Fragments that bound the antigen (binders) were identified at a frequency of 1/200; this relatively low frequency may be due to the presence of non-functional heavy and light chains in the Fab library. Ideally, amplification of the hybridoma RNA with specific 5′ heavy and light chains should generate Fab fragments that bind at a much higher frequency.

The DNA sequences of the binders were obtained to identify the clone that exactly represents the monoclonal catalytic antibody 43C9. Comparison of the light chain deduced N-terminal amino acid sequence of antibody 43C9 and the deduced amino acid sequences of ten of the binders indicated that five of the clones (8a11, 8a12, 8a1, 7a2, 7a4) had the correct light chain. Three of these clones (8a1, 8a11, 8a12) were identical and differed from each of the other two (7a2, 7a4) by a single amino acid in the framework region. All the clones had the same heavy chain sequence; comparison with the N-terminal sequence of the authentic antibody was not possible because its N-terminus is blocked.

Purified Fab from each of the ten clones described above was assayed for catalytic activity; 8a11, 7a2 and 7a1 hydrolysed amide **2** at a rate clearly above the background rate (Fig. 3). The reaction was inhibited completely by the addition of transition state analogue **3**, 20 μM. This indicated that the observed catalysis was exclusively due to the Fab. SDS-PAGE of the catalytic recombinant Fabs showed a single species at 50 kDa. Reducing conditions gave a doublet at 25 kDa, indicative of a single pure Fab fragment. Because of the limited amount of Fab produced in our system, detailed kinetic analysis has not been possible. Overexpression of the catalytic Fab is currently being sought, to facilitate the kinetic studies.

Discussion

The bacteriophage λ vector system developed for the expression of Fab fragments is ideally suited for studying the structure and mechanism of any desired monoclonal antibody. We have successfully expressed a monoclonal catalytic Fab in this system and have shown that it retains the ability to catalyse a specific amide hydrolytic reaction.

Future studies will be aimed at identifying more binders from the library which also display catalytic activity. The success of these will hinge upon our ability

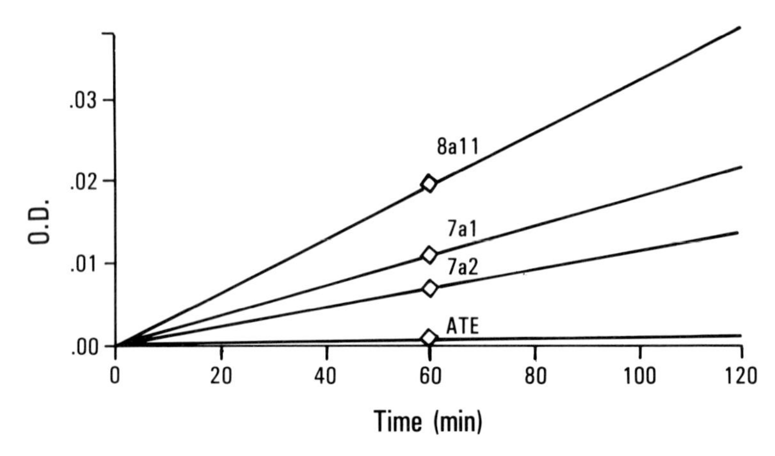

FIG. 3. Hydrolysis of carboxyamide **2** by Fab clones 8a11, 7a2 and 7a1. Hydrolysis was carried out at 37 °C with 2 μM antibody, 1 mM substrate in ATE buffer, pH 9.0. The differences in the observed rates seen for each clone probably reflect inaccuracies in protein concentration determination rather than clone differences. The background hydrolysis was measured with the substrate alone; in all cases the reaction was monitored at 405 nm.

to obtain a better system for expressing the protein, possibly utilizing Summer's baculovirus system (Smith et al 1983). More efficient screening for catalytic antibodies might be achieved via a genetic selection process.

Finally, a general solution to the antibody catalysis of a peptide bond may be obtained using the phage technology presented. Recently, we have constructed a single chain antibody with a coordination site for metals (Iverson et al 1990). When this site is incorporated into the light chain of an Fab fragment, a bound metal ion could act as a hydrolytic cofactor when properly aligned with a heavy chain which binds a small peptide sequence. The possibility of such a reaction appears remote; however, by taking advantage of the large numbers and combinations available through the combinatorial library, opportunities for success are within reach.

References

Chomczynski P, Sacchi N 1987 The calcium-sensitive dye arsenazo III inhibits calcium transport and ATP hydrolysis by the sarcoplasmic reticulum calcium pump. Anal Biochem 162:156–159

Huse WD, Sastry L, Iverson SA et al 1989 Expression of the antibody repertoire in phage λ. Science (Wash DC) 246:1275–1281

Iverson BL, Iverson SA, Roberts VA et al 1990 Metalloantibodies. Science (Wash DC) 249:659–662

Janda KD, Schloeder D, Benkovic SJ, Lerner RA 1988 Induction of an antibody that catalyses the hydrolysis of an amide bond. Science (Wash DC) 241:1188–1191

Lerner RA, Benkovic SJ 1988 Principles of antibody catalysis. Bioessays 9:107–112
Sanger F, Nicklen S, Coulson AR 1977 DNA sequencing with chain-terminating inhibitors. Proc Natl Acad Sci USA 74:5463–5467
Sastry L, Alting-Mees M, Huse WD et al 1989 Cloning of the immunological repertoire in *Escherichia coli* for generation of monoclonal catalytic antibodies: construction of a heavy chain variable region-specific cDNA library. Proc Natl Acad Sci USA 86:5728–5732
Short JM, Fernandez JM, Sorge JA, Huse WD 1988 Lambda ZAP: a bacteriophage lambda expression vector with in vivo excision properties. Nucleic Acids Res 16:7583–7600
Smith GE, Summers MD, Fraser MJ 1983 Production of human beta interferon in insect cells infected with a baculovirus expression vector. Mol Cell Biol 3:2156–2165

DISCUSSION

Hansen: The exact placing of an amino acid is critical for enzyme catalysis. Do you have a sense, perhaps from Sargeson's work (Buckingham et al 1970), of how precise one has to be in orienting a carbonyl group near the metal ion to see catalysis?

Lerner: I don't know. They were basically looking at an intramolecular situation, because the substrate was directly bound to an open site on the coordination complex.

Martin: Another structural detail to consider is the geometry of the antibody's metal-binding ligands. A slight difference in the relative positions of metal-binding amino acid side chains could have a dramatic effect on the catalytic efficiency of the antibody. The coordination geometry of metals in natural enzymes is often distorted: for example, the tetrahedral geometry of cobalt-substituted carboxypeptidase A is markedly irregular compared to simple tetrahedral complexes of cobalt such as cobalt tetrachloride. In their entatic state hypothesis, Vallee & Williams (1968) proposed that the distorted coordination geometry is a critical feature of metalloenzymes in that it causes the metal to be unusually reactive—in their terms 'poised for catalysis'.

Lerner: Isn't that flying in the face of results from a great number of coordination complex experiments?

Jencks: Ground state strain of that kind can change the properties of the ions and the ligands and the metal, certainly; but to say that the ground state strain or distortion directly influences the transition state is wrong. It may provide a system which has a proper pK or oxidation potential or whatever, that will lead to a transition state more readily, and this might be done better with another metal that has a different size and a different potential, but it doesn't relate directly to the stability of the transition state.

Martin: If the enzyme or antibody binds a metal with tetrahedral geometry, say by three amino acid side chains and a reactive water molecule, the effect of the distorted coordination geometry might be to fine tune the pK of the metal-bound water molecule, thereby making it more reactive.

Hansen: In the crystal structure of the metalloantibody (Iverson et al 1990), are there any peptide bonds of the antibody itself near the metal ion which might cause autodecomposition?

Lerner: We tried to do that experiment with copper(II). We think we have seen rather restricted proteolysis, but that uses other copper coordination sites outside the binding pocket of the antibody.

Plückthun: Have you looked at the effect of the presence of the metal ion on the folding of the protein?

Lerner: Yes. The protein can re-fold with and without the metal.

Plückthun: Are you surprised that the metal gets in and out so easily?

Lerner: No; it does so in enzymes. We were careful to do the whole experiment in metal-free water, because we didn't want any extraneous metals to get into the site.

Schultz: There is a phenolic oxygen fairly close to the metal ion in the antibody combining site. Have you looked at hydrolysis of fluorescein esters?

Lerner: No.

Iverson: We tried fluorescein diacetate and obtained no rate enhancement for the hydrolysis of this compound by the antibody in the presence or absence of the metal.

Lerner: But in terms of other partners for coordination, which Peter Schultz is asking about, there is another possibility, namely an aspartate that could be a coordination partner.

Suckling: Have you looked for electron transfer reactions in any metalloantibodies?

Lerner: Not yet, but we hope to.

Wilson: Have you tried a general screen with peptides to see whether any other antigens bind, as Scott & Smith (1990) did?

Lerner: No; it would be very interesting to do. Raymond Dwek's lab found a metal-binding site in a known antibody. If one is interested in accidents, one ought to screen catalytic antibodies, plus and minus metal. But metals can also make enzymes function considerably worse!

Jencks: I suspect you have thought of a good many things that metals will do, when they are attached to antibodies. Would you say any more on what you think about doing in the future?

Lerner: We plan to try to solve the so-called 'pre-monensin syndrome', which means how many chemist man hours are required to make monensin from pre-monensin, the problem being that there are so many chemically identical alkenes. We want to selectively epoxidize those molecules to do stereochemistry. We also, like Peter Schultz, want to try oxygen transfer reactions, particularly hydroxylation of steroids.

There are two important things to be done. One is to use these metalloantibodies for asymmetric inductions, where one would not have to use chiral auxiliaries because the antibody would give the system its asymmetry.

The success of that approach depends on (1) the binding specificity, and (2) how much trouble we shall have when we use relatively strong reagents like peracids.

Secondly, the single most important experiment one could do, thinking of the medical applications, is the selective cleavage of peptide bonds, using a recognition unit of seven amino acids rather than one or two. Proteases are relatively promiscuous enzymes and depend in part on the site environment rather than on the sequence of amino acids. One could not inject trypsin, for instance, into a person because it's too promiscuous. The ability to position ZnOH selectively in relation to seven amino acids would be tantamount to developing a restriction enzyme for protein molecules.

Benkovic: We also provide spectroscopists with a chance to investigate copper environments. There is a scarcity of models for copper(I) and copper(II) in various binding sites: by having a system that we shall be able to characterize by X-ray crystallographic analysis and manipulate at will, we should obtain interesting spin echo and electron paramagnetic resonance spectra for copper in these kinds of sites.

Lerner: If one is also interested in the flight of the electron in the protein cavity, there is no reason why one couldn't put two metals in there at defined distances and measure the rates of electron transfer.

Schultz: Could one use metal ion detoxification (such as HgII resistance) as the selection pressure to induce antibodies that have a metal-binding site?

Lerner: I think so. People have similar concepts to what you are talking about for immunization with alkylating agents: what if you were to immunize with a compound containing an open coordination site, such as HgII?

Hansen: The prospect of peptidases with more restricted recognition sites is very exciting. On a specific point, where is the metal ion in relation to the seven amino acid binding cleft that you mentioned?

Lerner: I don't think one can generalize about this. Because Ian Wilson is our neighbour, we have looked very closely at what we could do to cut a peptide in complexes whose structure he has solved. I don't believe a metal site would be close enough to any of the peptides that he has in his X-ray crystal structures to deliver the hydroxyl group effectively.

Hansen: The reverse question is whether the metal site would block peptide binding.

Lerner: If you look at the site in a 'cut-away' view, and if you assume that the peptide goes into it like a straight molecule, which it probably does not, it would not be easy to approximate a scissile bond, let's say next to a ZnOH. So you might wish to go further up the site. An antibody molecule often has a groove that could accommodate a peptide. I can imagine that the peptide will lie in that groove and approximate a scissile amide to the zinc ion. However, it remains to be shown that the present construction is good for peptide bond cleavage.

Hansen: Is the binding constant such that the metalloantibody holds together at zero metal concentration?

Lerner: The formation constant for CuII is micromolar. The constant for zinc is unknown.

Hansen: What about using a metalloantibody *in vivo*, therapeutically?

Lerner: At the moment one would have to pre-load. If the disease you want to cure is critical, you could infuse both the antibody and the zinc into the patient to increase the *in vivo* concentration of zinc. You could not do this with copper or other more toxic metals.

We hope soon to finish the coordination chemistry of the L1-L3 site, where we have tried thiols and carboxylates. I would be pleased to get a coordination constant of 10^8–10^9 there. There are many permutations of those three side chains that one needs to look at.

Harris: Many proteins have calcium associated with them. Have you thought about trying to introduce calcium-binding sites into antibodies?

Lerner: Steve Benkovic pointed out to me that when a protein binds carbohydrates it tends to use calcium. It's very difficult to make an antibody with a good binding constant for carbohydrate. Elvin Kabat said that he had never seen one better than millimolar. So to build an antibody that binds a carbohydrate, one might do as you suggest.

Harris: Calcium binding occurs via aspartate residues: where would be the best site for those to be placed?

Lerner: We are making 3-Asp and 3-Glu sites to see how those work.

Paul: You mentioned that R. Dwek's group had a monoclonal antibody with a metal-binding site, is it known where the metal is bound?

Janda: R. Dwek and co-workers found that the MOPC315 antibody bound lanthanides (Dower et al 1978). The antibody was pretty well characterized by Steve Dower (1978). He found that two carboxylates bound a lanthanide in addition to binding a *p*-nitrophenyl phosphate as a reporter molecule.

Burton: I wonder if you should make so much of this; lanthanides bind to a number of sites on antibodies, including one on Fc.

Lerner: It is not going to be difficult to build good coordination sites. What is going to be hard is approximating the metal to the substrate.

References

Buckingham DA, Foster DM, Sargeson AM 1970 Cobalt(III)-promoted hydrolysis of glycine amides. Intramolecular and intermolecular hydrolysis following the base hydrolysis of the *cis*-$[Co(en)_2Br(glyNR_1R_2)]^{2+}$ ions. J Am Chem Soc 92:6151–6158

Dower SK 1978 D Phil thesis, Worcester College, Oxford University, UK

Dower SK, Gettins P, Jackson R, Dwek RA, Givol D 1978 The binding of 2,4,6-trinitrophenyl derivatives to the mouse myeloma immunoglobulin A protein MOPC 315. Biochem J 169:179–188

Iverson BL, Iverson SA, Roberts VA et al 1990 Metalloantibodies. Science (Wash DC) 249:1275–1281
Scott JK, Smith GP 1990 Searching for peptide ligands with an epitope library. Science (Wash DC) 249:386–390
Vallee BL, Williams RJP 1968 Metalloenzymes: the entatic nature of their active sites. Proc Natl Acad Sci USA 59:498–505

Binding and multiple hydrolytic sites in epitopes recognized by catalytic anti-peptide antibodies

Sudhir Paul, Donald R. Johnson and Richard Massey*

*Departments of Pharmacology, Biochemistry and Pathology and Microbiology, University of Nebraska Medical Center, Omaha, NE 68198, and *IGEN Inc, Rockville, MD 20852, USA*

Abstract. Autoantibodies purified from humans catalyse the hydrolysis of the neurotransmitter, vasoactive intestinal peptide (VIP). Evidence that the hydrolysis of VIP is due to antibodies includes: the antibody preparations are free of detectable non-immunoglobulin (non-Ig) contamination; the hydrolytic activity is removed by precipitation with anti-human IgG antibody; human B lymphoblastoid cells transformed with Epstein–Barr virus secrete hydrolytic antibodies in culture; the Fab fragments of the antibodies exhibit VIP hydrolysis; and affinity chromatography on immobilized VIP permits purification of specific antibodies with greatly enriched hydrolytic and binding activities. One of the catalytic antibody preparations hydrolyses the Gln-16–Met-17 bond. Studies with synthetic VIP fragments showed that the epitope recognized by this antibody is formed by VIP(15–28). Important binding interactions are contributed by VIP(22–28), a sequence four residues distant from the scissile bond. Antibodies from a second subject hydrolyse six peptide bonds in VIP, clustered between residues 14 and 22. These bonds link amino acids of different charge, size and hydrophobicity, suggesting that the hydrolytic repertoire of the antibodies is considerable. The antibodies do not hydrolyse peptides unrelated in sequence to VIP. Cleavage of several peptide bonds in VIP by polyclonal antibody preparations may be due to several antibodies, each with a unique cleavage specificity. Alternatively, a single antibody may make catalytically productive contact at multiple peptide bonds in the substrate, because of conformational flexibility of VIP or of the antibody active site. Purified light chains from the catalytic antibodies hydrolysed VIP more rapidly than did intact antibodies. The residues constituting the catalytic site of an antibody may be encoded in germline V-region genes or may arise during maturation of the antibody response.

1991 Catalytic antibodies. Wiley, Chichester (Ciba Foundation Symposium 159) p 156–173

The ability of the immune system to synthesize antibodies capable of binding diverse non-self and self antigens specifically with high affinity derives from sequence variability in the complementarity-determining regions in the antibody active site. During maturation of the antibody response, sequence alterations in the active site may result in increased antigen binding affinity, which is thought to drive selective clonal expansion of cells expressing the antibody on their surface. Enzymic catalysis involves substrate binding, facilitation of bond breakage or formation by chemically reactive amino acids present in the enzyme active site (sometimes assisted by metal ions), and dissociation of products. Since the complementarity-determining regions of antibodies are prone to sequence alteration (French et al 1989), catalytic antibodies could arise during the normal development of the immune response, by fortuitous, correctly oriented placement of chemically reactive amino acids in the antibody active site. Current thinking is dominated by the belief that the sole biological function of antibody active sites is to bind antigens. Although a systematic search for catalysis by antibodies to natural antigens has not been performed, there are several reports that immunization with the ground state of haptens provokes formation of antibodies that accelerate chemical reactions (Raso & Stollar 1975, Kohen et al 1980a,b, Sucking et al, this volume). Observations summarized in this paper indicate that, at least in one case, the immune system mounts a catalytic antibody response to a polypeptide antigen, vasoactive intestinal peptide (VIP). In principle, catalysis by antibodies to natural antigens would be advantageous in that it could render immunological defence against microbes and tumours more efficient. It remains to be determined whether mechanisms exist to ensure selection of catalytic antibodies over antibodies that bind antigens stoichiometrically.

Considerable progress has been made in manipulating antibody active sites to produce catalytic activity. However, efficient catalysis of energetically unfavourable reactions, such as peptide bond cleavage, by such antibodies has yet to be accomplished. Antibodies capable of peptide bond cleavage may eventually be useful in research and therapy, either in their native form or after active site manipulations that generate new catalytic specificities.

Biology of vasoactive intestinal peptide

This peptide is a 28 amino acid neurotransmitter (Fig. 1) homologous in sequence to peptide histidine isoleucine, growth hormone releasing factor, secretin and glucagon (Mutt 1988). Originally isolated from the intestine, VIP is now recognized to be widely distributed in the central and peripheral nervous systems (Said 1982). It is implicated in the regulation of blood pressure, bronchial tone, neuroendocrine activity, exocrine secretion, immune regulation and growth of tumour and non-tumour cells (Said 1982, 1984, Brenneman et al 1990, Prysor-Jones et al 1989).

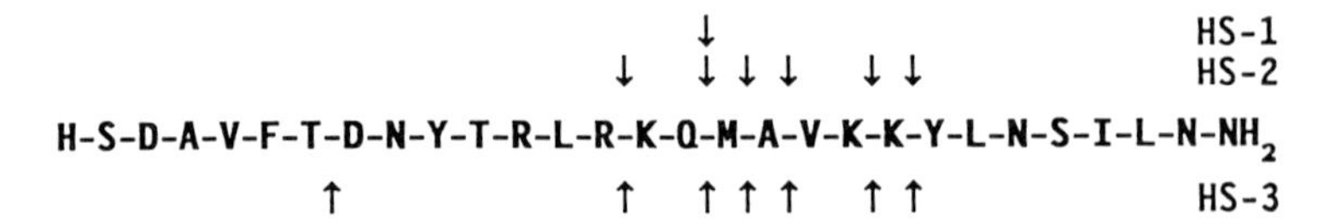

FIG. 1. The amino acid sequence of vasoactive intestinal peptide showing the peptide bonds cleaved by polyclonal antibodies (arrows) from three humans (HS-1, HS-2 and HS-3).

Human autoantibodies to VIP

The blood of some asthma patients and healthy humans with histories of muscular exercise contains VIP-binding autoantibodies (Bloom et al 1979, Paul et al 1985, 1989a, Paul & Said 1988). These are specific for VIP, judged by their poor reaction with peptides related to VIP; their binding constants for VIP are in the 10^7–$10^9\,M^{-1}$ range. The antibodies found in asthma patients bind VIP with greater affinity than do those from healthy subjects. VIP is a potent bronchodilator and antibodies against it may be linked to the pathophysiology of asthma. These autoantibodies occur naturally but the antigenic stimulus leading to their formation is not known: candidates include viral determinants similar in sequence to VIP, such as peptide T, an epitope found on the AIDS virus (Ruff et al 1987); dietary ingestion of avian and ichthyic VIP known to be structurally different from human VIP (Nilsson 1975, Thwaites et al 1989); and muscular exercise (Paul & Said 1988).

Identification of autoantibodies that cleave VIP

Immunoglobulin G molecules (IgG) purified from the blood of nine humans positive for VIP-binding antibodies were assayed for their ability to hydrolyse [^{125}I]Tyr-VIP, estimated as the amount of radioactivity rendered soluble in trichloroacetic acid (Paul et al 1989b). IgG from three subjects (designated HS-1, HS-2 and HS-3) were positive and several lines of evidence demonstrated that the hydrolysis of VIP was mediated by antibodies (Paul 1989, Paul et al 1989b, 1990a,b, D. J. Volle, M. Sun, S. Paul, unpublished observations 1990).

1. No contaminating non-immunoglobulin proteins were detected by denaturing gel electrophoresis and silver staining. The IgG preparations were composed of two protein bands of 61 kDa and 26 kDa, stainable by antibodies to human heavy and light chains, respectively.
2. After purification of the hydrolytic IgG by ammonium sulphate precipitation and DEAE-cellulose chromatography, the hydrolytic activity was recovered quantitatively in the fraction bound to immobilized protein G, a bacterial protein that binds the Fc region of IgG. Precipitation with anti-human IgG antibody removed 70–90% of [^{125}I]Tyr-VIP hydrolytic activity. The residual activity is probably due to incomplete precipitation of the antibody.

3. Elution of the hydrolytic activity was coincident with the peak of IgG from a high performance gel filtration column.
4. Digestion of the IgG with papain reduced the mass of the hydrolytic activity to 56 kDa, as estimated by SDS-electrophoresis and gel filtration: this is the size of Fab fragments.
5. To remove potential protease contaminants non-covalently associated with the antibodies, we subjected IgG and Fab to repeated ultrafiltration on 100 kDa and 30 kDa cut-off filters. This procedure did not diminish their hydrolytic activity. Similarly, the hydrolytic activity remained associated with antibodies during gel filtration at pH 2.7, arguing strongly against a non-covalently associated conventional protease.
6. Specific antibodies prepared by affinity chromatography of IgG on immobilized VIP exhibited enriched VIP hydrolytic activity [HS-1, 2076-fold; HS-2, 1154-fold; and HS-3, 100-fold]. The profiles of VIP hydrolytic activity and VIP binding activity recovered during elution of the antibodies with an acidic buffer from the affinity column were the same.
7. The agent responsible for VIP hydrolysis exhibits K_m values indicative of tighter substrate binding than by conventional proteases (see below).
8. B lymphoblastoid cells from HS-1 and HS-3 prepared by transformation with Epstein–Barr virus (EBV) were found to secrete VIP hydrolytic antibodies in culture.

Catalytic efficiency

The hydrolytic activities of the purified IgG and the affinity-purified antibodies were saturable with increasing VIP concentration. Both types of antibody preparations displayed Michaelis–Menten saturation kinetics. Scatchard plots for VIP binding by the antibodies were linear ($r = 0.97–0.99$). Since the saturation analyses of binding and hydrolysis indicated antibodies with a single K_d (HS-1, 0.4 nM; HS-2, 29 nM) and K_m (HS-1, 38 nM; HS-3, 2.2 μM), it is likely that the interaction with VIP was predominantly due to a single species of antibody in each preparation. Compared to classical proteases, the antibodies turn over slowly ($k_{cat} = 0.4–15.6\ min^{-1}$). However, the antibodies bind VIP tightly and consequently their kinetic efficiencies (k_{cat}/K_m) approach those of proteases (HS-1, $4.1 \times 10^8\ min^{-1}\ M^{-1}$; HS-3, $2 \times 10^5\ min^{-1}\ M^{-1}$).

Scissile bonds

To locate the scissile bonds, we incubated unlabelled VIP with IgG purified by protein G–Sepharose chromatography or specific antibodies purified by chromatography on immobilized VIP. The resultant fragments of VIP were separated by reverse-phase high performance liquid chromatography (HPLC) and identified by N-terminal sequencing. In control incubations, VIP was treated with non-immune IgG or assay diluent. Antibodies from one subject (HS-1) produced two VIP fragments, VIP(1–16) and VIP(17–28). We concluded that HS-1 antibodies cleave VIP at a single peptide bond, the Gln-16–Met-17 bond (Fig. 1) (Paul et al 1989b). Multiple VIP fragments were isolated after exposure of VIP to antibodies from HS-2 and HS-3 (S. Paul and D. J. Volle, unpublished 1990). The scissile bonds were identified on the basis of detection of peptide fragments that: (i) contained an N-terminal residue other than the N-terminal His of full-length VIP, (ii) possessed a clearly identifiable amino acid sequence five or more residues in length that was identical to internal subsequences of VIP, and (iii) were present at a level of 10 pmoles or more in each sequencing cycle. From these studies, six peptide bonds were identified that were cleaved by both HS-2 and HS-3 antibodies. These six bonds are clustered in the segment of VIP spanning residues 14–22. A seventh scissile bond located between residues 7 and 8 was cleaved only by HS-3 antibodies. From the amounts of fragments recovered, cleavage of VIP by HS-2 antibodies was most pronounced at the Lys-21–Tyr-22 bond, and by HS-3 antibodies at the Ala-18–Val-19 bond. The amino acid residues at the seven scissile bonds differ in charge, size and hydrophobicity. The P_1 site can be occupied by Thr, Arg, Gln, Met, Ala or Lys, and the P_1' site by Asp, Lys, Met, Ala, Val or Tyr. Thus, the hydrolytic repertoire of the antibodies must be considerable.

Peptide bond hydrolysis in the absence of a catalyst usually requires exposure to harsh conditions (e.g. acid pH and elevated temperatures). However, Asn–X bonds can undergo slow spontaneous cleavage at physiological pH and temperature via intramolecular nucleophilic attack on the peptide bond carbonyl by the amide side chain; the half-life value for this deamidative cleavage process is of the order of months (Geiger & Clarke 1987). Theoretically, Gln-X bonds could undergo analogous deamidative cleavage. The Gln-16–Met-17 bond belongs to this class of peptide bonds, but the other six bonds cleaved by our antibodies do not. In view of the diversity of residues at the scissile bonds, the potential destabilization of peptide bonds by the side chains of certain amino acids is not sufficient to explain antibody-mediated cleavage of VIP. The free energy of antigen binding by antibodies can be as large as 50–75 kJ/mole (corresponding to K_a 10^9–10^{13} M^{-1}). Anti-hapten antibodies produced by immunization against transition state analogues are capable of catalysing acyl transfer reactions, including hydrolysis of esters and substituted amides (Tramontano et al 1986, Janda et al 1988). Binding of antibodies to the transition state presumably provides sufficient energy to drive these reactions. Peptide bonds, however, are stable bonds and their hydrolysis probably requires more energy than would be available just from binding of the transition state to the antibody. This conclusion is supported by observations that IgG preparations from six humans bound VIP but failed to hydrolyse the peptide. Although the mechanism of the hydrolysis of VIP remains to be defined, the antibodies probably act like conventional proteases, with chemically reactive amino acids in the catalytic site of the antibody (e.g. His, Ser, Asp) participating directly in peptide bond hydrolysis. Direct evidence for this proposition is available from preliminary studies in which hydrolysis of VIP by antibodies from HS-1 was inhibited by diisopropylfluorophosphate, a serine protease inhibitor, and that by antibodies from HS-3, by EDTA, a metalloprotease inhibitor.

The antibodies from HS-2 isolated by affinity chromatography on immobilized VIP cleaved six peptide bonds in VIP. One possibility is that the cleavage of VIP is mediated by six different antibodies, each with a unique hydrolytic specificity. The Scatchard plot of VIP binding and the Lineweaver-Burk plot of VIP hydrolysis by the antibodies were essentially linear. If several antibodies are present, they exhibit experimentally indistinguishable K_d and K_m values. The alternative is cleavage of VIP at multiple sites by a single antibody with a polyspecific or non-specific hydrolytic activity. Conformational flexibility in VIP or the antibody active site could permit catalytically productive contact at multiple peptide bonds. This hypothesis is suggested by the following considerations: (i) all six bonds cleaved by this antibody preparation are clustered in a 9-residue sequence, VIP(14–22), (ii) the slow antibody turnover suggests a catalytic site that is not optimized for rapid positioning and cleavage of a specific substrate bond, (iii) in aqueous solutions, VIP displays little structural order (Bodanszky & Bodanszky 1982, Robinson et al 1982), and (iv) conformational isomerism of an antibody may permit multispecific antigen binding (Stevens et al 1988). The binding and hydrolytic specificities of an antibody may not be due to interaction with the same residues or subsequence of the substrate (see section on Site-specificity below). A polyspecific or non-specific hydrolytic activity need not compromise the substrate specificity of an antibody, since the binding step alone could be sufficient to confer specificity. The catalytic antibodies we have examined fulfilled the expectation of substrate specificity.

Substrate specificity

We investigated the ability of our antibodies to hydrolyse three peptides unrelated in sequence to VIP. The antibodies from HS-2 and HS-3 failed to hydrolyse ^{125}I-insulin, ^{125}I-insulin-like growth factor II and ^{125}I-atrial natriuretic peptide, as judged by the amount of radioactivity recovered as intact peptide by HPLC or TCA-precipitation after treatment with the antibodies or assay diluent. Under experimental conditions identical to those used for these peptides, 38.0–53.7% of [^{125}I]Tyr-VIP was hydrolysed.

Site-specificity

Antibodies from HS-1 cleave VIP at a single peptide bond (Gln-16–Met-17). Ten synthetic fragments, corresponding to linear subsequences of VIP, were

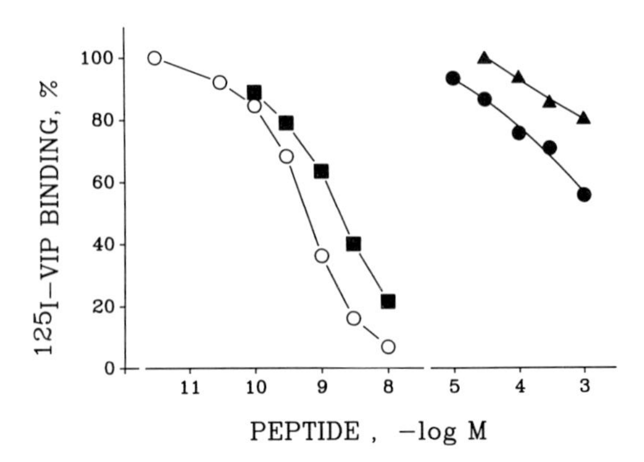

FIG. 2. Saturable binding of radiolabelled vasoactive intestinal peptide (VIP) to antibodies from HS-1: from competitive inhibition by VIP(1–28) (○), VIP(15–28) (■), VIP(22–28) (●) and VIP(18–24) (▲).

examined for reactivity with this antibody preparation (Paul et al 1990b). Hydrolysis of [^{125}I]Tyr-VIP was undetectable under the experimental conditions utilized for these binding assays. A large C-terminal subsequence, VIP(15–28), was bound with high affinity ($K_i = 1.25$ nM), suggesting that it forms the antibody-binding epitope (Fig. 2). Of seven heptapeptide and one octapeptide subsequences examined, only VIP(22–28) and VIP(18–24) inhibited binding of [^{125}I]Tyr-VIP. Both these sequences are distant from the scissile bond. The HS-1 antibodies did not show detectable binding of short VIP

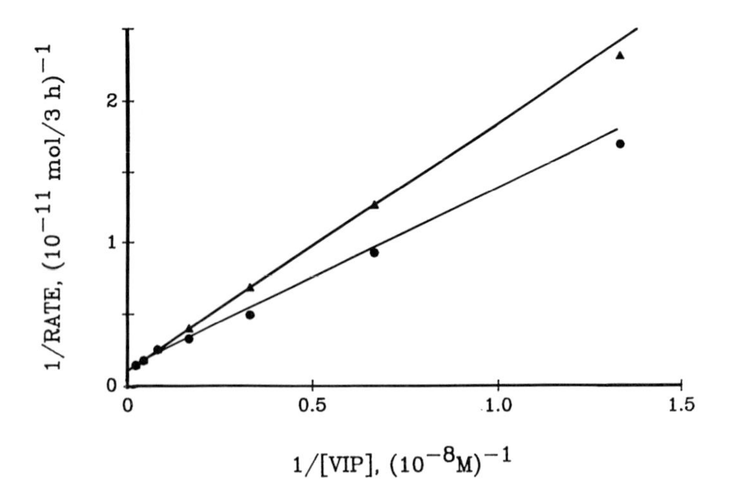

FIG. 3. Catalytic hydrolysis of VIP(1–28) by antibodies from HS-1 in the absence (●) and presence (▲) of 100 μM VIP(22–28).

subsequences that encompass the scissile bond, namely VIP(11–17), VIP(15–21) and VIP(13–20). Competition for binding by VIP(22–28) was approximately eightfold greater than that by VIP(18–24), suggesting that the reactivity of the latter peptide is due to the shared sequence VIP(22–24). The binding energy of VIP(22–28) (19.3 kJ/mol) was 41% that for VIP(15–28). Hydrolysis of VIP was inhibited competitively by VIP(22–28) (Fig. 3). K_i for VIP(22–28) calculated from these data was 260 μM. The reactivity of VIP(22–28) with the catalytic autoantibody appears to be sequence specific, since: (i) this peptide inhibited [^{125}I]Tyr-VIP binding and hydrolysis competitively, (ii) the inhibition was observed in the presence of an excess of an unrelated protein, bovine serum albumin, and (iii) VIP(1–7), a non-binding fragment, did not inhibit hydrolysis of VIP. VIP(22–28) may, therefore, be considered a 'subepitope' that contributes some, but not all, antibody binding interactions (Fig. 4).

Antigen-binding sites in antibodies are relatively large, capable of contacting as many as 15–22 amino acids of antigen (Amit et al 1986, Davies et al 1990). Antibody interactions with short peptide fragments are unlikely to mimic exactly the interaction with an epitope present on intact VIP. For example, residues 15–21 may contribute non-specifically to the binding energy of VIP(22–28), since N-terminal extension of small peptides with irrelevant amino acids or octanoyl groups is known to increase their antibody binding affinity (Benjamini et al 1968a,b). This possibility is consistent with studies (Novotny et al 1989) suggesting that a subset of residues within an epitope may contribute most of the binding energy in an antigen–antibody complex. A second alternative is that VIP(22–28) may be the major component of a conformational epitope, but one

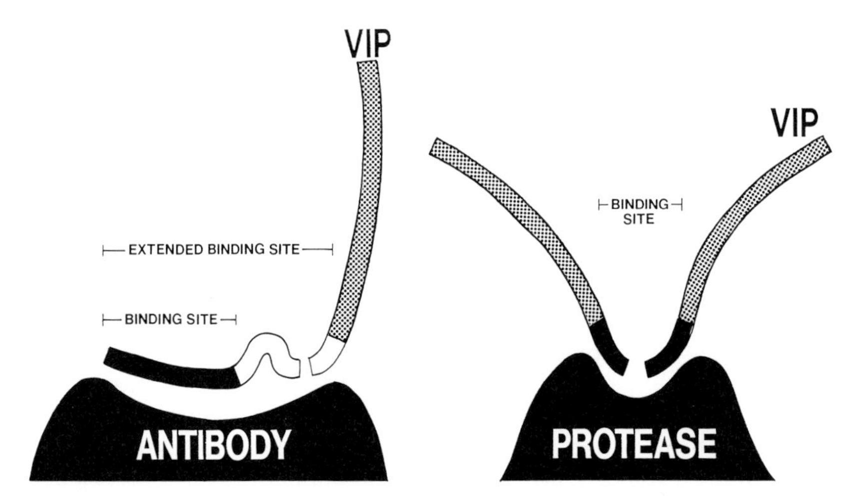

FIG. 4. The catalytic antibody from HS-1 (left) binds a seven-amino acid subsequence of VIP distant from the cutting site (Gln-16–Met-17, shown as a gap). A conventional protease (right) recognizes only a short stretch of amino acids at the site to be cut.

or more of residues 15–21 may be integral to this epitope. Although linear VIP fragments consisting of residues 15–21, 11–17 and 13–20 did not bind the antibodies, the residues at the scissile bond (Gln-16–Met-17) must contact the catalytic group(s) in the antibody active site. Thus whether Gln-16, Met-17 and their flanking residues make important contributions to binding remains unclear.

Conventional peptidases hydrolyse a broad range of proteins with specificities dictated primarily by the type of amino acids at or in the immediate vicinity of the scissile bond (Barrett 1986). In contrast, the catalytic antibodies from HS-1 exhibit significant binding of VIP(22–28), although four residues are present between this subsequence and the Gln-16–Met-17 bond that is hydrolysed. The recognition of a distant peptide sequence is likely to underlie the tight binding and high level of substrate specificity of this catalytic antibody, compared to conventional peptidases.

Catalysis by antibody light chains

Catalytic residues in the antibody active site could be the same ones that bind antigens; if the two types of residues are different, they are probably situated close to each other. The catalytic and binding residues need not be present on the same chain. In recent studies (S. Paul and M. Sun, unpublished), we have purified the light chains of VIP-specific antibodies from HS-1 to apparent homogeneity. The purified light chains hydrolysed VIP with a turnover number greater than the intact, parent antibodies, and also with increased K_m. This suggests that the dissociated light chains have reduced antigen binding affinity, but are capable of rapid catalysis. It appears that as the antibody is progressively freed from structural constraints imposed by the disulphide-bonded, four-chain architecture of the intact IgG molecule, its enzymic character increases, probably at the expense of antigen recognition specificity.

A perspective

Fragments of mature protein antigens derived by antibody-mediated hydrolysis are likely to exhibit diminished biological activity. Antibodies capable of catalytic hydrolysis of proteins could serve as efficient and specific mediators of immunological defence against microbes and tumours, and, conversely, of the pathophysiology of autoimmune disorders. The findings summarized here provide the first, and, as yet, the only example of catalytic cleavage of peptide bonds in an antigen by naturally occurring antibodies. Although the generality of catalytic antibodies is yet to be examined, Nature usually repeats itself. Viewed teleologically, the ability to generate and select mutations that confer catalytic activity on antibodies directed against viral, bacterial and tumour antigens could be advantageous in immunological defence. It remains to be determined whether sequence information for catalytic antibodies is present in germline V-region genes.

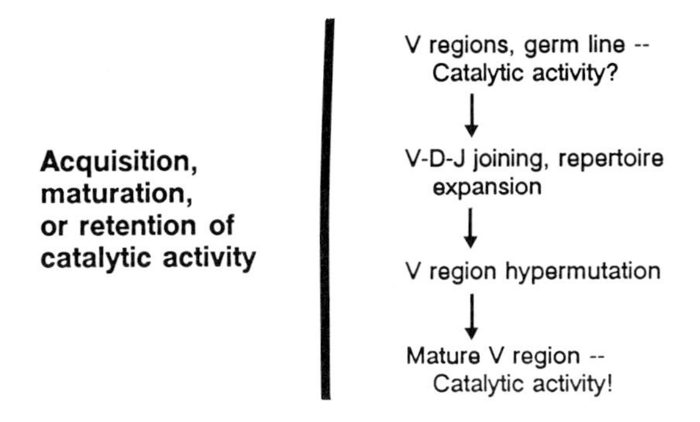

FIG. 5. When does an antibody become catalytic?

During maturation of the antibody response, catalytic antibodies could arise spontaneously by the same mechanisms that are responsible for antibody V-region sequence variability, i.e. inaccuracies in V-D-J joining, somatic hypermutation, and heavy and light chain combinatorial variability (Fig. 5). Randomly arising catalytic antibodies could be mere epiphenomena unless mechanisms exist to ensure selection of catalytic antibody-producing lymphocytes. It is believed that during immunization with an antigen, lymphocytes that express antibodies with the highest binding affinity on their surface bind antigen preferentially and are driven into clonal proliferation. The antibodies to VIP exhibit tight binding, suggesting that they are a result of an active selection process. One possibility is that the continuous presence of antigen bound to a surface antibody leads to desensitization of lymphocytes, as happens in most other receptor–ligand systems. A hypothetical mechanism that might permit selection of catalytic antibodies over non-catalytic antibodies is the escape of catalytic antibody-producing lymphocytes from desensitization owing to cleavage and release of surface bound antigen.

Acknowledgements

Supported by the National Institutes of Health, U.S.A. We thank Steve Eklund, Tony Rees and George Whitesides for suggestions and criticism, and Deanna Volle and Sun Mei for experimental assistance.

References

Amit AG, Mariuzza RA, Phillips SEV, Poljak RJ 1986 Three-dimensional structure of an antigen–antibody complex at 6Å resolution. Nature (Lond) 313:156–158

Barrett AJ 1986 An introduction to proteinases. In: Barrett AJ, Salvesen G (eds) Proteinase inhibitors. Elsevier, Amsterdam, p 3–22

Benjamini E, Shimizu M, Young JD, Leung CY 1968a Immunochemical studies on the tobacco mosaic virus protein. VI. Characterization of antibody populations following immunization with tobacco mosaic virus protein. Biochemistry 7:1253–1260

Benjamini E, Shimizu M, Young JD, Leung CY 1968b Immunochemical studies on the tobacco mosaic virus protein. VII. The binding of octanoylated pepetides of the tobacco mosaic virus protein with antibodies to the whole protein. Biochemistry 7:1261–1264

Bloom SR, Barnes AJ, Adrian TE, Polak JM 1979 Autoimmunity in diabetics induced by hormonal contaminants of insulin. Lancet 1:14–17

Bodanszky M, Bodanszky A 1982 Synthesis of VIP and related peptides: structure–activity relationships. In: Said SI (ed) Vasoactive intestinal peptide. Raven Press, New York, p 11–22

Brenneman DE, Nicol T, Warren D, Bowers LM 1990 Vasoactive intestinal peptide: a neurotrophic releasing agent and an astroglial mitogen. J Neurosci Res 25: 386–394

Davies DR, Padlan EA, Sheriff S 1990 Antibody–antigen complexes. Annu Rev Biochem 59:439–474

French DL, Laskow R, Scharff MD 1989 The role of somatic hypermutation in the generation of antibody diversity. Science (Wash DC) 244:1152–1157

Geiger T, Clarke S 1987 Deamidation, isomerization and racemization at asparaginyl and aspartyl residues in peptides. Succinimide linked reactions that contribute to protein degradation. J Biol Chem 262:785–794

Janda KD, Schloeder D, Benkovic SJ, Lerner RA 1988 Induction of an antibody that catalyzes the hydrolysis of an amide bond. Science (Wash DC) 241:1188–1191

Kohen F, Kim JB, Barnard G, Linder HR 1980a Antibody-enhanced hydrolysis of steroid esters. Biochim Biophys Acta 629:328–337

Kohen F, Kim JB, Linder HR, Eshhar Z, Green B 1980b Monoclonal immunoglobulin G augments hydrolysis of an ester of the homologous hapten. FEBS (Fed Eur Biochem Soc) Lett 111:427–431

Mutt V 1988 VIP and related peptides. Ann NY Acad Sci 527:1–19

Nilsson A 1975 Structure of the vasoactive intestinal peptide from chicken intestine. The amino acid sequence. FEBS (Fed Eur Biochem Soc) Lett 60:322–326

Novotny J, Bruccoleri RE, Saul FA 1989 On the attribution of binding energy in antigen–antibody complexes McPC 603, D1.3, and HyHel-5. Biochemistry 28:4735–4749

Paul S 1989 A new effector mechanism for antibodies: catalytic cleavage of peptide bonds. Cold Spring Harbor Symp Quant Biol 54:283–286

Paul S, Said SI 1988 Human autoantibody to vasoactive intestinal peptide: increased incidence in muscular exercise. Life Sci 43:1079–1084

Paul S, Heinz-Erian P, Said SI 1985 Autoantibody to vasoactive intestinal peptide in human circulation. Biochem Biophys Res Commun 130:479–485

Paul S, Said SI, Thompson A et al 1989a Characterization of autoantibodies to VIP in asthma. J Neuroimmunol 23:133–142

Paul S, Volle DJ, Beach CM, Johnson DR, Powell MJ, Massey RJ 1989b Catalytic hydrolysis of vasoactive intestinal peptide by human autoantibody. Science (Wash DC) 244:1158–1162

Paul S, Volle DJ, Mei S 1990a Affinity chromatography of catalytic autoantibody to vasoactive intestinal peptide. J Immunol 145:1196–1199

Paul S, Volle DJ, Powell MJ, Massey RJ 1990b Site-specificity of a catalytic vasoactive intestinal peptide antibody: an inhibitory VIP subsequence distant from the scissile peptide bond. J Biol Chem 265:11910–11913

Prysor-Jones RA, Silverlight JJ, Jenkins JS 1989 Oestradiol, vasoactive intestinal peptide and fibroblast growth factor in the growth of human tumour cells *in vitro*. J Endocrinol 120:171–177

Raso V, Stollar BD 1975 The antibody–enzyme analogy. Comparison of enzymes and antibodies specific for phosphopyridoxyltyrosine. Biochemistry 14:591–599

Robinson RM, Blakeney EW, Mattice WL 1982 Lipid-induced conformational changes in glucagon, secretin, and vasoactive intestinal peptide. Biopolymers 21:1217–1228
Ruff MR, Martin BM, Ginns EI, Farrar WL, Pert CB 1987 CD4 receptor binding peptides that block HIV infectivity cause human monocyte chemotaxis. FEBS (Fed Eur Biochem Soc) Lett 211:17–22
Said SI (ed) 1982 Vasoactive intestinal peptide. Raven Press, New York
Said SI 1984 Vasoactive intestinal peptide (VIP). Current status. Peptides (NY) 5:143–150
Stevens FJ, Chang C-H, Schiffer M 1988 Dual conformations of an immunoglobulin light-chain dimer: heterogeneity of antigen specificity and idiotope profile may result from multiple variable-domain interaction mechanisms. Proc Natl Acad Sci USA 85:6895–6899
Suckling C, Tedford CM, Proctor GR, Khalaf AI, Bence LM, Stimson WH 1991 Catalytic antibodies: a new window on protein chemistry. In: Catalytic antibodies. Wiley, Chichester (Ciba Found Symp 159) p 201–210
Thwaites DT, Young J, Thorndyke MC, Dimaline R 1989 The isolation and chemical characterization of a novel vasoactive intestinal peptide-related peptide from a teleost fish, the cod, *Gadus morhua*. Biochim Biophys Acta 999:217–220
Tramontano A, Janda KD, Lerner RA 1986 Catalytic antibodies. Science (Wash DC) 234:1566–1570

DISCUSSION

Lerner: Using polyclonal serum is not a good experiment because you have no idea of the concentration of the antibody. What is worse, you have no idea of your substrate concentration because there are so many antibodies present that bind the antigen but have no catalytic activity.

Paul: The chemical nature of an antibody is identical whether it is produced *in vitro* by mouse hybridomas or secreted into the blood *in vivo* to form part of a polyclonal mixture. Therefore, there is no reason why polyclonal antibody preparations should not possess catalytic activity, given sufficient quantity of the responsible antibody. One can quantitate catalytic antibodies on the basis of their binding a substrate or inhibitor (Paul et al 1990b, Benkovic et al 1988). If non-catalytic antibodies are present, as in polyclonal sera, they may decrease substrate availability, which would lead to underestimation of the strength of binding and kinetic efficiency of the catalytic antibody.

Benkovic: An important aspect of our work is the purity of the antibody species—that we have a single entity which we can examine in detail.

Suckling: We looked at the polyclonal serum against our antigen (see Suckling et al, this volume) to work out analytical techniques and we did find catalytic activity. We found more catalysis once we had the monoclonal antibodies.

Schultz: We have looked at four different reactions with both polyclonal and monoclonal antibodies: in each case the catalytic activity of the polyclonal preparation turned out to be an enzyme contaminant. These activities included deaminases, phosphodiesterases, peptidases and glycosidases.

For the phosphodiesterase, IgMs were generated against a vanadate derivative of uridine. Affinity columns, sizing columns and ion exchange columns were then used to separate ribonucleases from the antibody. Polyacrylamide gels showed the antibody to be pure and catalytic activity was inhibited by the hapten. However, a sequencing gel of the RNA cleavage pattern showed that the sequence specificity of the purified IgM activity was identical to that of ribonuclease. This rules out a catalytic antibody, unless there is an amazing coincidence.

So when looking at any enzyme activities that are present in sera, it is crucial to be very careful. In general, cationic exchange columns are necessary after protein A purification to remove contaminating glycosidases and peptidases from an antibody.

Hilvert: Have you ever looked at abzyme activity on a non-denaturing gel?

Schultz: We are attempting such experiments using pH-sensitive gel overlays.

Paul: There are two issues: contamination of antibodies with other antibodies and contamination of antibodies with enzymes. Ascites fluid, the usual source of monoclonals, contains many esterases, proteases and ribonucleases that could co-purify with antibodies, as does plasma, the source of polyclonals.

Lerner: There is no reason to work with polyclonal antibodies any more. Polyclonal antibodies are merely a very large collection of monoclonal antibodies. You just don't want to do an experiment with an unknown mixture.

Paul: The reason we work with polyclonal antibodies is that they can catalyse efficient, sequence-specific cleavage of peptide bonds (Paul et al 1990a): this has not yet been demonstrated for a monoclonal antibody. Secondly, study of the role of catalytic antibodies in physiology and disease will often require the use of polyclonal clinical specimens.

Schultz: I see one major objection to this. If Steve has a monoclonal catalytic antibody and I have Steve's cell line, I can purify the same antibody and reproduce his results. If you are dealing with polyclonal antisera, you run into problems with reproducibility of results. That may be the strongest argument for using monoclonal antibodies in the catalytic antibody field.

Paul: A fitting analogy is the proof that antibodies bind antigens. Monoclonal antibodies were not available and those studies were done with polyclonals. We are at an early stage with respect to catalytic cleavage of peptide bonds by antibodies, which is why our group uses polyclonals. We are sharing our peptide bond-cleaving antibodies with scientists at IGEN and would be happy to provide them to others who are interested.

Benkovic: I certainly wouldn't select, for polyclonal assay, reactions for which there are known enzymes, because the chance is far too high of having a trace contaminant that gives what appears to be a high antibody turnover number.

Harris: Dr Paul, after you used papain to generate the Fab region, what steps did you take to remove the enzyme? And did you check the cleavage pattern of VIP after hydrolysis by the Fab fragments?

Paul: We used immobilized papain and removed any enzyme that may have leached off the column by gel filtration. Fab fragments prepared from non-catalytic IgG do not hydrolyse VIP. Preliminary experiments suggest that the Fab fragments and IgG have the same cleavage specificity.

Jencks: Do you know the pH dependence of the hydrolytic activity?

Paul: pH 8 appears to be optimal for all three antibody preparations that we have looked at. The pH dependence shows a sharp maximum and low activity at the two extremes, which were pH 5 and pH 10. Activity falls dramatically between pH 9 and 10.

Benkovic: Do all the products appear at the same rate, for the two antibody preparations that cleaved the peptide substrate at different places?

Paul: No. Antibodies from subject HS-3 cleave most efficiently at Ala-18–Val-19; those from HS-2 mainly at Lys-21–Tyr-22.

Benkovic: So k_{cat} is an average and the K_m is a composite?

Paul: The K_m is indeed a composite value. Model peptides in which only a single bond is cleaved will be needed to make more precise measurements. We are working with polyclonal antibodies, so we don't know if it is one antibody cleaving multiple peptide bonds or several antibodies, each with a unique cleavage specificity. We are trying to raise monoclonal antibodies to address this issue.

Benkovic: Have you looked at which bond is hydrolysed by the light chains alone?

Paul: We have not. The K_m for the light chains is 4.1 mM, so they have not lost antigen recognition completely. I would expect an isolated light chain to be less specific than the original antibody.

Benkovic: What did you measure to show the light chains were reactive?

Paul: We measured total product—the degraded fragments which were not precipitated by TCA.

Wilson: Have you done any amino acid sequencing of the Fab fragments or the intact IgG?

Paul: We are trying to do this. In preliminary experiments with small amounts of affinity-purified antibodies we were unable to determine any sequence, perhaps because of blocked N-termini.

Kang: Are you attempting to clone the genes encoding the IgG molecule from the EBV-transformed B lymphoblastoid cells?

Paul: Yes. The cells produced antibodies for a while, then the cells died. We are wondering whether to use growth factors to maintain cell viability, to hybridize the cells with myelomas or simply to use PCR primers to amplify the entire repertoire present in the polyclonal cultures and clone the catalytic antibody cDNA by expression screening.

Iverson: Have you re-bled the asthma patients, to see if the catalytic activity is reproduced in preparations from the same patient?

Paul: In one healthy subject catalytic antibodies were found in four different samples taken over 10 months. The same bond was cleaved by the antibody preparation from two of these samples. The K_m of the antibody preparations was different—38 nM and 112 nM. k_{cat} for each was 15.6 and 6.5 min^{-1}, respectively (Paul et al 1990a). A bronchitic patient analysed for catalytic antibodies in different bleeds gave similar results.

Schultz: What concentration of antibody are you using in the catalytic assays?

Paul: It varies. Affinity-purified antibodies are used at concentrations of 1–10 nM. IgG preparations are used at micromolar concentrations because the antibodies that catalyse VIP constitute only a fraction of the sample.

Leatherbarrow: Your Ig from HS-1 that is inhibited by diisopropyl-fluorophosphate (DFP) could be used for active site titration. You could investigate whether the amount of inhibitor reacting is equal to the total amount of antibody present, or whether the amount of reaction is consistent with a serine protease contaminant. Have you done that experiment?

Paul: We have not done active site titration. Our data cannot be explained adequately by the presence of a serine protease or other conventional enzymic contaminant, as I showed in my paper. I would like to link DFP to the antibody covalently to determine whether the catalytic site is in the light chain or the heavy chain.

Leatherbarrow: If you can show that you get a stoichiometric reaction with DFP, and that gives a stoichiometric loss of activity, then you could rule out contamination.

Paul: We may do that experiment, but I don't think it will give us the definitive answer you predict. In any purified enzyme preparation, only a fraction of the molecules may exhibit the desired activity (Dunn 1989): some may be denatured, others may be in the wrong conformation. So the catalyst concentration determined by active site titration may not agree with antibody concentration determined using another method. Also, a protease contaminant bound covalently to an antibody would elude the distinction you are trying to make.

Leatherbarrow: It would not have the same molecular weight.

Paul: It could be a very small peptide, an unconventional protease contaminant, although this is a far-fetched argument. The specificity of the antibody from HS-1 is unlike that of any known protease (Paul et al 1990b). Moreover, this antibody has nanomolar K_m values, whereas those of proteases are millimolar.

Lerner: When you published the first experiment on this subject, I thought that you had a high spontaneous background rate of VIP hydrolysis. Now it seems there is no spontaneous hydrolysis.

Paul: We use iodinated VIP as the substrate in our routine assays. Iodinated VIP is prepared once every 45 days or so, and purified by HPLC. Despite storage at −80 °C under the best conditions with protectants present, we see spontaneous breakdown in the freezer. Thus when we start a hydrolysis assay only about 90% of the radioactivity is TCA precipitable. This is a standard finding with iodinated proteins. Over the three hours of the hydrolysis assay there is no further breakdown of the peptide.

Lerner: Do you always follow the reaction by a count of radioactive iodine?

Paul: No. Usually, when we want qualitative answers, for example the identity of the scissile bonds, we use unlabelled VIP as substrate: there is no radiolabelled VIP present. Peter Schultz has suggested that radiolysis could be a factor in antibody-catalysed hydrolysis of VIP (Shokat & Schultz 1990); that clearly is not the case.

Plückthun: Have you incubated your radioactive peptide with control sera over very long times, to see whether you get breakdown and whether you obtain a specific break? If there is a contaminating protease, one would assume that this is shared by many human beings, and one way of detecting it would be to see what fragments are obtained after extensive incubations in sera from normal people. I am concerned whether likely contaminants, namely proteases present in human serum, will preferentially attack the same peptide bonds as the hypothetical catalytic antibody.

Paul: We have not done the experiment with serum proteases. George Caughey has described cleavage of VIP by mast cell chymase and tryptase at Arg-14–Lys-15, Lys-20–Lys-21 and Tyr-22–Leu-23 (Caughey et al 1989). Five of the seven antibody-sensitive bonds in VIP are apparently not cleaved by these enzymes.

Shokat: When you retain specific catalytic activity of the antibody preparations after purification on a VIP–Sepharose column, what degree of purification have you achieved?

Paul: With different antibodies we see between 100- and 2100-fold purification after chromatography on VIP–Sepharose.

Shokat: Does the specific activity increase linearly at each step?

Paul: Yes. There is a linear increase in the amount of hydrolysis with increasing IgG and purified antibody preparations, provided that the substrate concentration is not limiting.

Shokat: How did you measure the K_d?

Paul: This was done by a binding assay at 4 °C under conditions that do not permit hydrolysis of the peptide (Paul et al 1989). At relatively high salt concentrations the antibodies still bind VIP but do not hydrolyse it.

Shokat: So the comparison of K_m and K_d surely can't be made?

Paul: In the assay we use to measure K_d, the interaction between VIP and the antibody is arrested at the binding step. A direct comparison of K_m and K_d is difficult because the assay conditions for binding and hydrolysis are different. The K_d measurement shows we are dealing with a high affinity antibody directed against the ground state of the substrate. We can also calculate antibody concentrations from the binding assays.

Janda: You use a VIP affinity column. Does the antibody not catalyse the hydrolysis of that column?

Paul: No, because the column is run in high salt conditions which prevent hydrolysis by the antibody. The purified antibodies are then dialysed before being assayed. Also, IgG from healthy people includes an unidentified inhibitor that must be removed by dialysis or ultrafiltration before antibody-catalysed hydrolysis can occur. Using non-hydrolytic IgG without salt as the starting material, we see some acquisition of catalytic activity in the affinity-purified antibodies even without dialysis. The inhibitor removed by dialysis and affinity chromatography could be non-specific or directed towards the active site, which would be particularly interesting.

Iverson: One would expect affinity purification to eliminate an active site binding inhibitor. If you have VIP on the column, it should displace any bound inhibitor. After elution of antibody, there should not be any inhibitor left bound to the antibody binding site.

Paul: If the inhibitor occupies exactly the same site as VIP, yes. The alternative is that extensive washing of the immobilized antibody on the VIP–Sepharose column removes an inhibitor that is bound at a site distant from the cleavage site. We also see activation of non-hydrolytic antibodies when they are immobilized on a protein G column and washed extensively. We are still uncertain about the nature of the inhibitor.

Janda: You proposed a mechanism for this antibody whereby the loss of ammonia occurred via a cyclic-type intermediate formed by the peptide backbone (Paul et al 1989). One way to test for that would be the release of ammonia. But you are now suggesting that metals are involved in the catalysis by one set of antibodies (those from HS-3).

Paul: Antibodies have hypervariable active sites and different antibodies are likely to use different catalytic mechanisms. For example, the metal-stimulated antibodies from HS-3 are not inhibited by DFP, and those from HS-1 are sensitive to DFP but not to EDTA. It is possible that cyclization and deamidation (Geiger & Clarke 1987) play a role in the cleavage of the Gln–Met bond. They cannot be the mechanism by which the other bonds in VIP are cleaved by our antibodies because those bonds do not undergo deamidative cleavage.

Benkovic: How do you explain the fact that for one of the substrates an analogous substrate sequence apparently does not inhibit the antibody? Why does the antibody suddenly appear to have lost ligand-binding ability?

Another problem is that the k_{cat} for the light chain catalyst differs by only a factor of 10 from that of trypsin. Thus, a light chain alone, without any heavy chain recognition, is almost as good as a highly evolved protease. That is too remarkable, surely?

Paul: It is remarkable only because it was not anticipated. I never think of Nature as 'too remarkable'. Antibodies are not conventional enzymes: the specificity profile I have described appears to involve important binding interactions at sites in the substrate distant from the scissile bond (Paul et al 1990b). The light chain has a faster rate of turnover than the intact antibody:

it is efficient, but not as good as trypsin. Our working hypothesis is that there is a generic domain in the light chain that has good catalytic activity. I am pleased with this because we are seeking efficient antibody catalysts.

Benkovic: Can you inhibit the generic catalyst with a substrate-like sequence?

Paul: We haven't tested different sequences from VIP as substrates for hydrolysis by the light chain.

Benkovic: The usual antibody recognition sequence is only 6–7 residues: how can residues outside that inhibit?

Paul: Structural studies show that between 15 and 22 residues of an antigen can contact the antibody active site (Davies et al 1990). A hexapeptide or heptapeptide may be the minimum recognition sequence for an antibody, but the high energy of binding to larger polypeptides appears to involve extensive contact of the antibody with the substrate. Within an epitope composed of 15–22 amino acids, strong binding interactions appear to occur at a few sites. VIP(22–28) looks like a sequence that contributes to binding even though it is distant from the scissile bond.

References

Benkovic SJ, Napper AD, Lerner RA 1988 Catalysis of a stereospecific bimolecular amide synthesis by an antibody. J Am Chem Soc 85:5355–5358

Caughey GH, Leidig F, Viro NF, Nadel JA 1989 Substance P and vasoactive intestinal peptide degradation by mast cell tryptase and chymase. J Pharmacol Exp Ther 241:133–137

Davies DR, Padlan EA, Sheriff S 1990 Antibody–antigen complexes. Annu Rev Biochem 59:439–474

Dunn BM 1989 Determination of protease mechanism. In: Beynon RJ, Bond JS (eds) Proteolytic enzymes, a practical approach. IRL Press, Oxford, p57–82

Geiger T, Clarke S 1987 Deamidation, isomerization and racemization at asparaginyl and aspartyl residues in peptides. Succinimide linked reactions that contribute to protein degradation. J Biol Chem 262:785–794

Paul S, Volle DJ, Beach CM, Johnson DR, Powell MJ, Massey RJ 1989 Catalytic hydrolysis of vasoactive intestinal peptide by human autoantibody. Science (Wash DC) 244:1158–1162

Paul S, Volle DJ, Powell MJ, Massey RJ 1990a Site specificity of a catalytic vasoactive intestinal peptide antibody: an inhibitory VIP subsequence distant from the scissile peptide bond. J Biol Chem 365:11910–11913

Paul S, Volle DJ, Mei S 1990b Affinity chromatography of catalytic autoantibody to vasoactive intestinal peptide. J Immunol 145:1196–1199

Shokat KM, Schultz PG 1990 Catalytic antibodies. Annu Rev Immunol 8:335–363

Suckling CJ, Tedford TM, Proctor GR, Khalaf AI, Bence LM, Stimson WH 1991 Catalytic antibodies: a new window on protein chemistry. In: Catalytic antibodies. Wiley, Chichester (Ciba Found Symp 159) p 201–210

Antibody catalysis of carbon-carbon bond formation

Donald Hilvert

Departments of Chemistry and Molecular Biology, Research Institute of Scripps Clinic, 10666 North Torrey Pines Road, La Jolla, CA 92037, USA

Abstract. We have used rationally designed transition state analogues to generate antibodies that catalyse two important carbon–carbon bond forming reactions: a bimolecular Diels–Alder cycloaddition and a unimolecular Claisen rearrangement. Our tailored immunoglobulin catalysts (abzymes) exhibit all the properties of naturally occurring enzymes, including substantial rate accelerations, substrate specificity, and high regio- and stereoselectivity. As first generation abzymes are generally inferior to naturally occurring enzymes, we are also employing classical genetic selection strategies to augment their chemical efficiency. We have expressed the antibody that catalyses the Claisen rearrangement of chorismate in yeast cells that lack natural chorismate mutase activity. Improved versions of the abzyme will be identified, following random mutagenesis, by their ability to repair this metabolic defect. The development and study of highly efficient catalytic antibodies promises to advance our understanding of how enzymes work and evolve, how protein function correlates with structure, and how entirely new enzymic activities can be created for use in research, industry and medicine.

1991 Catalytic antibodies. Wiley, Chichester (Ciba Foundation Symposium 159) p 174–187

By current estimates, the immune system of mammals, humans included, has the potential to generate at least *100 million* different antibodies, each with a distinctive binding site. This capacity exists to protect the body against disease-causing microorganisms and toxins. Recently, however, it has become possible to direct the diversity and specificity of this receptor-generating system toward the production of antibodies with tailored catalytic activities.

The preparation of enzyme-like antibodies, or abzymes, follows an original suggestion by Jencks (1969). The three-step procedure entails synthesizing stable compounds that mimic the transition state of the reaction to be catalysed, eliciting an immune response against such substances, and characterizing the monoclonal antibodies that are generated. If the transition state analogue is well designed, a significant fraction of the isolated antibodies will possess the desired catalytic activity.

In my laboratory, we have used rationally designed transition state analogues to create abzymes for two important carbon–carbon bond forming reactions: a bimolecular Diels–Alder cycloaddition and a unimolecular Claisen rearrangement (Hilvert et al 1988, 1989). These concerted pericyclic processes are sensitive to strain and proximity effects and have minimal requirements for chemical catalysis. Detailed study of these systems can consequently contribute substantially to our understanding of the importance of these factors in enzymic catalysis. In addition, cycloadditions and sigmatropic rearrangements are among the most valuable transformations available to organic chemists for constructing complex natural products and synthetic materials of all kinds (Desimoni et al 1983). Tailored antibody catalysts able to steer these reactions down single regio- and stereochemical pathways may therefore have great practical value.

Diels–Alder cycloadditions

The Diels-Alder reaction involves concerted addition of a conjugated diene to an alkene to generate a cyclohexene derivative (Sauer 1966, 1967). Two carbon–carbon sigma bonds are formed at the expense of two π bonds, and both regio- and stereochemistry are rigorously controlled in the course of the reaction. The transition structure of this [4+2] pericyclic process is highly ordered with a boat-like cyclohexene conformation.

Although 'Diels-Alderase' enzymes have never been found in Nature, there is no fundamental reason why cycloadditions cannot be catalysed by proteins. The activation barrier for bimolecular Diels-Alder reactions has a very large entropic component, typically in the range -30 to $-40\,\text{cal}\,K^{-1}\,\text{mol}^{-1}$ (Sauer 1966). By preorganizing the reactants through favourable binding interactions, an antibody should be able to minimize this tremendous loss of translational and rotational entropy (Page & Jencks 1971, Jencks 1975).

We have focused initially on the Diels-Alder reaction between tetrachlorothiophene dioxide and *N*-alkylmaleimides (Fig. 1). The process was originally investigated by Raasch (1980) and shown to proceed via an unstable bicyclic molecule **1**. This compound undergoes a rapid chelotropic elimination to produce a dihydrophthalimide derivative, which, under aqueous reaction conditions, is subsequently oxidized to the phthalimide. The transition structures for both the initial cycloaddition and the subsequent cycloreversion resemble the high energy intermediate **1**, in which a boat-like conformation is forced on the cyclohexene ring by the SO_2 bridge. Consequently, we felt that stable bicyclic compounds that resemble **1**, like the hexachloronorbornene derivative **2**, might be suitable haptens for eliciting an antibody combining site with a topology appropriate for bringing the tetrachlorothiophene dioxide and maleimide molecules together. We anticipated that antibodies raised against **2** would have an important additional advantage: because **2** does not closely resemble the planar phthalimide adduct, product inhition—which is normally

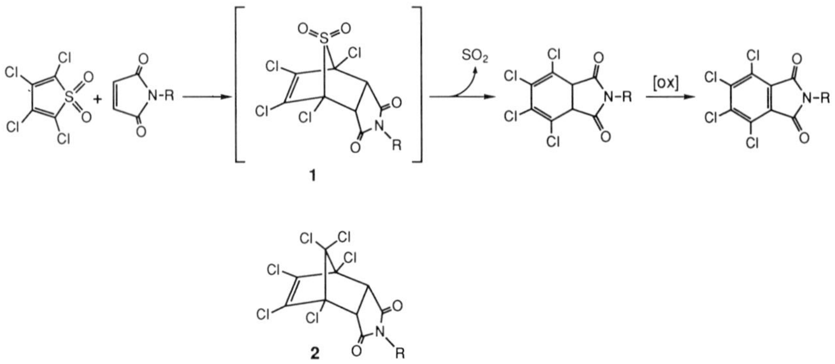

FIG. 1. Diels–Alder cycloaddition between tetrachlorothiophene dioxide and *N*-alkylmaleimides. A protein conjugate of the transition state analogue **2** was used to generate a catalytic antibody for this process.

a serious concern for protein-catalysed bimolecular reactions—should not be a problem.

An immune response was invoked against a conjugate of **2** and the carrier protein keyhole limpet haemocyanin (KLH), and monoclonal antibodies were prepared using standard protocols (Hilvert et al 1989). Immunoglobulins that bound the transition state analogue tightly were identified by ELISA, propagated in mouse ascites, and purified to homogeneity. Prior to kinetic evaluation the antibodies were subjected to exhaustive reductive methylation with formaldehyde and sodium cyanoborohydride to prevent an undesired side reaction between substrate and the ϵ-amino groups of lysine residues on the antibody.

Several of the permethylated immunoglobulin G antibodies that bind the transition state analogue **2** also accelerate the two-step reaction between tetrachlorothiophene dioxide and *N*-ethylmaleimide (Hilvert et al 1989). In the presence of the most efficient antibody, 1E9, the reaction proceeds cleanly with multiple turnovers. *N*-Ethyl-tetrachlorophthalimide was isolated from the reaction mixture, characterized, and shown by competition ELISA to bind more than 10^3-times less tightly to the combining site than hapten **2**. Sulphur dioxide, a by-product of the cycloreversion, was detected by its ability to bleach malachite green. It is present in less than stoichiometric amounts, because it mediates oxidation of the initially formed dihydrophthalimide.

The accelerated cycloaddition occurs within the induced antibody binding pocket, as judged by the fact that the reaction is strongly inhibited by the hexachloronorbornene hapten. Mixing the antibody with one equivalent of hapten per binding site prevents the reaction, even in the presence of a thousandfold excess of substrate. Furthermore, substrate specificity correlates precisely with hapten structure. *N*-Alkylmaleimides with long alkyl side chains that mimic the linker in the KLH–**2** conjugate are better substrates for the abzyme than maleimides with short side chains, although the uncatalysed reaction rate is independent of maleimide structure.

Saturation kinetics, indicating a kinetic mechanism in which the antibody and substrates form a Michaelis-type complex prior to reaction, is also observed. The apparent kinetic parameters $k_{cat} = 4.3 \pm 0.3\ min^{-1}$ and $(K_m)_{NEM} = 21 \pm 4\ nM$ were obtained at pH 6 and 25 °C by varying the concentration of *N*-ethylmaleimide and holding the concentration of tetrachlorothiophene dioxide constant at 0.61 mM (Hilvert et al 1989). Comparison of $(k_{cat})_{app}$ with the second order rate constant for the uncatalysed reaction ($k_{uncat} = 0.04\ M^{-1}\ min^{-1}$) gives an apparent effective molarity of 110 M. The effective molarity is the concentration of substrate that would be needed in the uncatalysed reaction to achieve the same rate as achieved in the antibody ternary complex. The observed value is several orders of magnitude larger than the maximum physically accessible concentration of substrate in aqueous buffer, therefore the antibody binding site confers a significant kinetic advantage on the bimolecular Diels–Alder reaction. An effective molarity of 110 M, however, represents only a lower estimate of the abzyme's true chemical efficiency, because low solubility prevented saturation of the antibody with diene. Under the conditions of the experiment $(k_{cat})_{app}$ is linearly dependent on thiophene dioxide concentration, so the true effective molarity must be substantially (at least 10 to 100 times) larger.

Catalysis of a Diels–Alder cycloaddition with an antibody demonstrates the feasibility of exploiting proximity effects to achieve substantial rate accelerations. These experiments are also notable for the use of chemical modification to extend the properties of the abzyme and for the strategy, implicit in the antigen design, of avoiding product inhibition. Given the importance of cycloadditions in organic synthesis, extension of these ideas to other systems is desirable. Diels–Alder reactions that occur poorly at ambient temperatures and pressure, or with low regio- and stereoselectivity, will be especially challenging targets.

Sigmatropic rearrangements

Sigmatropic processes constitute a second class of pericyclic reactions of great importance in organic synthesis (Desimoni et al 1983). In the unimolecular Claisen rearrangement of allyl enol ethers, for instance, a carbon–carbon bond is formed at the expense of a carbon–oxygen bond (Ziegler 1988). This concerted reaction proceeds via a transition structure with cyclicly delocalized electrons and a high degree of stereochemical control.

The conversion of (−)-chorismate into prephenate (Fig. 2) is a biologically relevant Claisen rearrangement (Weiss & Edwards 1980). It is accelerated 2×10^6 times over the uncatalysed thermal process by the enzyme chorismate mutase (EC 5.4.99.5). Extensive inhibitor studies (Andrews et al 1977, Chao & Berchtold 1982) and elegant stereochemical investigations (Sogo et al 1984)

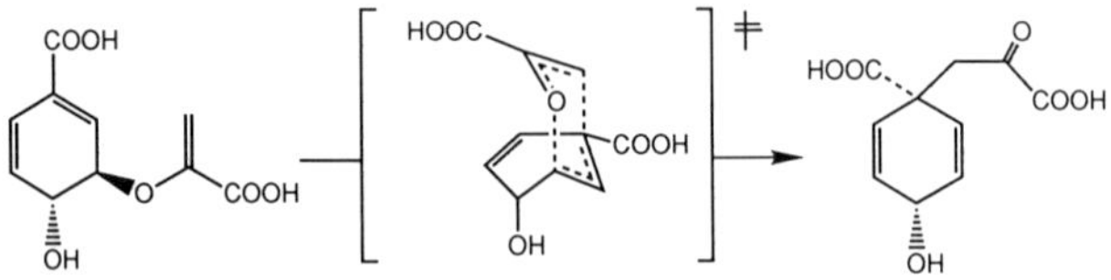

FIG. 2. The Claisen rearrangement of chorismate into prephenate. The endo oxabicyclic diacid **4** mimics the transition state structure **3** and was used to prepare antibodies with chorismate mutase activity.

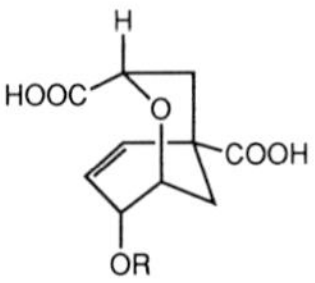

have suggested that the enzymic reaction proceeds through a conformationally restricted transition state with pseudo diaxial chair-like geometry **3**. Consistent with this idea, the endo oxabicyclic diacid **4**, designed by Bartlett and co-workers (1988) to mimic the putative high energy species, is currently the best inhibitor of chorismate mutase. Significantly, **4** has also proved effective in generating chorismate mutase abzymes.

We prepared and purified 45 monoclonal antibodies against **4** by standard techniques (Hilvert et al 1988). These were tested individually, spectroscopically and by high pressure liquid chromatography, for their ability to accelerate the rearrangement of chorismate into prephenate. Two of the antibodies, 1F7 and 27G5, significantly enhanced the rate of this process. We showed that prephenate was the authentic product of the antibody-catalysed reaction, and that its rate of formation equalled the rate of chorismate disappearance. As expected if rearrangement occurs in the induced binding pocket, the transition state analogue is a competitive inhibitor of catalysis with a K_i of about 0.6 μM. The catalysed reaction is also highly enantioselective: the antibody promotes rearrangement of only (−)-chorismate (Hilvert & Nared 1988). We exploited the stereoselectivity of this system to obtain optically pure (+)-chorismate by kinetic resolution of racemic substrate.

Saturation kinetics is observed for the antibody-catalysed chorismate mutase reaction, consistent with a kinetic scheme in which substrate and antibody form a Michaelis complex prior to the rearrangement event. Fitting the data obtained in aqueous buffer (pH 7.5, 25 °C) by the standard Michaelis–Menten treatment, we calculated values for the kinetic parameters k_{cat} and K_m of 0.072 min^{-1} and 51 μM, respectively (Hilvert et al 1988). Here, k_{cat} is the first order rate constant for interconversion of the bound substrate into the bound product.

It can be compared directly with the first order rate constant for the uncatalysed thermal rearrangement ($3.8 \times 10^{-4}\,min^{-1}$) to obtain a measure of the chemical efficiency of the abzyme. By this criterion, antibody 1F7 accelerates the reaction roughly 200-fold, in accord with the relative affinities of the antibody for chorismate and the oxabicyclic inhibitor. Although the observed rate enhancement is substantial, 1F7 is still 10^4 times less active than naturally occurring chorismate mutase from *Aerobacter aerogenes*.

Using the same hapten **4**, Jackson et al (1988) prepared a chorismate mutase abzyme that enhanced the rate of the 3,3-sigmatropic rearrangement by a factor of roughly 10^4. Their catalyst is 50 to 100 times faster than 1F7 and possesses 0.1% of the activity of the natural enzyme. This result demonstrates that non-ideal transition state analogues can be employed to obtain abzyme catalysts with quite respectable activities. The fact that the same antigen can induce several antibodies with different catalytic efficiencies underscores the importance of screening the immune system widely in order to identify those antibodies with optimal specific activities.

Extensive screening is also important for identifying antibodies with altered substrate specificities and stereoselectivities. Although a racemic hapten was used, the only chorismate mutase abzymes found in the two fusions studied accelerated the rearrangement of (−)-chorismate but not (+)-chorismate. Nevertheless, catalysts with opposite stereoselectivity could presumably be found by screening larger libraries of antibodies raised against the oxabicyclic diacid. Indeed, Janda et al (1989) have employed a racemic phosphonate ester to induce antibodies that hydrolyse either enantiomer of the analogous ester substrate.

We are currently dissecting the active site of our chorismate mutase antibody by a combination of chemical and genetic techniques. Attempts are also being made to obtain crystals suitable for X-ray diffraction studies. Information gained in such investigations will contribute substantially to our understanding of structure–function relationships in proteins. We expect that the design principles that produced the chorismate mutase abzymes will be extended and generalized to prepare antibody catalysts for sigmatropic processes that have no physiological counterpart. In addition to other Claisen-type rearrangements, the energetically demanding Cope rearrangement (especially the oxyanionic version), [2,3]-sigmatropic processes, and hydrogen atom shifts are attractive targets for catalysis.

Future directions

While the feasibility of generating antibody catalysts has been firmly established, this technology has yet to be applied to important practical problems. All first generation abzymes are orders of magnitude less active than their naturally occurring counterpart enzymes. Even those antibodies, like our Diels–Alderase catalyst, that promote non-physiological reactions do not work at optimal chemical efficiency. General strategies for improving these molecules are therefore needed, if the full potential of catalytic antibody technology is to be realized.

Naturally occurring enzymes have been brought to peak efficiency through many millions of years of evolution. We want to mimic this molecular selection process in the laboratory to augment the chemical efficiency of first generation abzymes. This could be accomplished, in principle, by expressing the antibody inside a microorganism and growing vast numbers of the transformants under conditions such that only those cells that produce an improved version of the catalyst survive. Recent progress in producing antibodies in microorganisms such as *Escherichia coli* and yeast makes this approach possible.

The chorismate mutase system is an ideal model to evaluate this strategy. The mutase is at the branchpoint in the biosynthetic pathway for production of aromatic amino acids in lower organisms (Weiss & Edwards 1980). Yeast cells that lack chorismate mutase, but have all the other enzymes in the shikimate pathway, cannot grow unless they receive phenylalanine and tyrosine in their diets (Jones & Fink 1982). Correction of this metabolic defect by overexpression of the chorismate mutase abzyme in the cytoplasm of the mutant yeast would provide a powerful assay system for identifying genetic changes that result in improved chemical efficiency.

In collaboration with Dr. James Hicks we have cloned and sequenced the genes encoding the heavy and light chains of the chorismate mutase antibody 1F7 and have engineered them for expression as an Fab protein in yeast cytoplasm (Bowdish et al 1991). The signal sequences for the heavy and light chains were removed and the heavy chain was truncated at the hinge region. The resulting fragments were cloned into a yeast expression vector behind an inducible, bidirectional promoter from the yeast GAL1–GAL10 gene complex. The resulting plasmid (Fig. 3) also contains the yeast 2 μm origin of replication and the *leu2d* selectable marker.

We transformed a strain of *Saccharomyces cerevisiae* yeast lacking natural chorismate mutase with this expression plasmid. Northern analysis of the transformants verified that large amounts of mRNA corresponding to the heavy and light chain genes of 1F7 are produced on induction with galactose. Gene transcription is not observed when the cells are grown on glucose rather than galactose. More importantly, the mRNA is translated efficiently into functional Fab protein, as judged by an ELISA. Under optimized conditions, overexpressed 1F7 Fab makes up roughly 0.1% of total cellular protein.

Antibody was isolated from yeast and purified by affinity chromatography on an anti-Fab antibody column. The protein had a relative molecular mass of 50 000 as judged by polyacrylamide gel electrophoresis under non-reducing conditions. The molecule is a disulphide-bonded dimer; in the presence of thiol the interchain disulphide is cleaved to yield the separated heavy and light chains ($M_r \approx 25\,000$). We have shown that the purified yeast Fab is also functional.

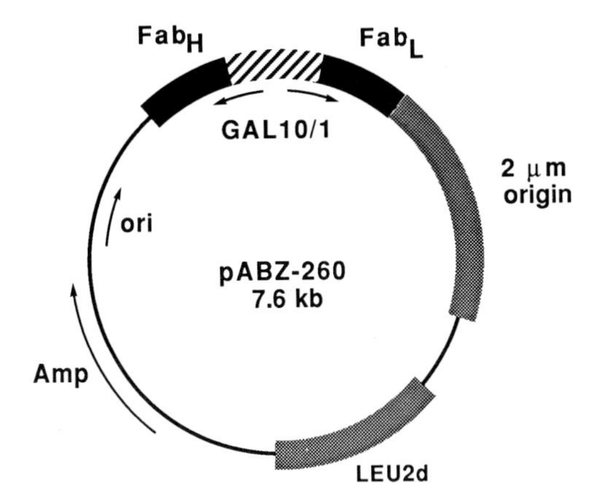

FIG. 3. Plasmid used to coexpress the heavy and light chains of 1F7 Fab intracellularly in the yeast *Saccharomyces cerevisiae*.

It binds the transition state analogue **4** and possesses the same specific activity in *in vitro* kinetic assays as the antibody isolated from hybridoma cells (Bowdish et al 1991).

The availability of a convenient expression system for the abzyme will allow us to probe the role of specific active site residues by site-directed mutagenesis. We can now test, for example, the hypothesis that positively charged lysines and arginines in the H2 and H3 complementarity-determining regions participate in catalysis by binding and orienting the negatively charged substrate carboxyl groups of chorismate.

Our primary goal, however, is to utilize the yeast expression system for genetic selection. Preliminary experiments show that the intracellular chorismate mutase activity is not enough to repair the genetic deficiency of the yeast host cells at the current level of Fab expression. Given that the abzyme is only 0.01% as active as the natural enzyme, this result is not surprising: for complementation, a more active antibody is needed. Toward that end, we have treated the antibody-containing yeast cells with ethyl methanesulphonate and have identified about 200 mutants that are able to grow in the absence of added tyrosine (Y. Tang, J. B. Hicks, D. Hilvert unpublished work 1990). Current studies will show whether the expressed antibody is responsible for the observed phenotype. We are also randomly mutagenizing the antibody-encoding genes directly, to augment catalytic activity even further.

If these experiments are successful, we will ultimately generate a family of related antibody molecules of increasing chemical efficiency. In essence, we will be monitoring the evolution of the abzyme in a living organism. By carrying out detailed chemical, kinetic and structural characterization of these 'evolutionary snapshots', we will be able to deduce the general rules that relate protein structure to protein function for the chorismate mutase system. We anticipate that molecular genetics will eventually provide a powerful and general tool for probing and improving the activity of any catalytic antibody.

Concluding remarks

The development of catalytic antibodies represents an exciting advance in the field of enzyme engineering. Abzymes exhibit all the features of true enzymes, including rate accelerations, substrate specificity, and regio- and stereoselectivity. Because the properties of these catalysts correlate precisely with hapten structure, they can be predetermined by the researcher through appropriate design. Consequently, tailored antibody catalysts have great promise as tools for studying how natural enzymes work, and as practical agents for accelerating reactions for which natural enzymes are unavailable, unstable, or otherwise unsuitable. Through the combination of chemistry and immunology with molecular biology and genetics, the catalytic antibody approach to enzyme design can be made even more powerful, and abzymes that rival naturally occurring enzymes may soon become readily available for a wide variety of practical applications.

Note added in proof

We have recently obtained compelling evidence that one of the EMS-treated, chorismate mutase-deficient yeast strains requires the activity of the intracellularly expressed 1F7 abzyme for growth under selective conditions (Y. Tang, J. B. Hicks, D. Hilvert, submitted). The key observations are: (1) segregation of the Phe^+ phenotype with the antibody-encoding plasmid; (2) correlation of the selective growth advantage in the absence of phenylalanine with induction of antibody gene transcription; and (3) reconstitution of the defective shikimate pathway by serial retransformation of the permissive yeast strain with the genes coding for the chorismate mutase abzyme, but not with those for an unrelated esterolytic antibody. These experiments demonstrate that catalytic antibodies can function effectively *in vivo* and suggest exciting applications for these molecules in the regulation of cellular function, alteration of cellular metabolism, and the destruction of carcinogens or other toxins.

Acknowledgements

This work was supported in part by grants from the National Institutes of Health (GM38273), the National Science Foundation (DMB-8912068) and a Faculty Research Award from the American Cancer Society.

References

Andrews PR, Cain EN, Rizzardo E, Smith GD 1977 Use of chorismate mutase inhibitors to define the transition state structure. Biochemistry 16:4848–4852

Bartlett PA, Nakagawa Y, Johnson CR, Reich SH, Luis A 1988 Chorismate mutase inhibitors: synthesis and evaluation of some potential transition-state analogues. J Org Chem 53:3195–3210
Bowdish K, Tang Y, Hicks JB, Hilvert D 1991 Yeast expression of a catalytic antibody with chorismate mutase activity. J Biol Chem, in press
Chao HSI, Berchtold GA 1982 Inhibition of chorismate mutase activity of chorismate mutase-prephenate dehydrogenase from *Aerobacter aerogenes*. Biochemistry 21: 2778–2781
Desimoni G, Tacconi G, Barco A, Pollini GP 1983 Natural products synthesis through pericyclic reactions. ACS Monograph 180. American Chemical Society, Washington, DC
Hilvert D, Nared KD 1988 Stereospecific Claisen rearrangement catalyzed by an antibody. J Am Chem Soc 110:5593–5594
Hilvert D, Carpenter SH, Nared KD, Auditor MTM 1988 Catalysis of concerted reactions by antibodies: the Claisen rearrangement. Proc Natl Acad Sci USA 85:4953–4955
Hilvert D, Hill KW, Nared KD, Auditor MTM 1989 Antibody catalysis of the Diels–Alder reaction. J Am Chem Soc 111:9261–9262
Jackson DY, Jacobs JW, Sugasawara R, Reich SH, Bartlett PA, Schultz PG 1988 An antibody-catalyzed Claisen rearrangement. J Am Chem Soc 110:4841–4842
Janda KD, Benkovic SJ, Lerner RA 1989 Catalytic antibodies with lipase activity and R and S substrate selectivity. Science (Wash DC) 244:437–440
Jencks WP 1969 Catalysis in chemistry and enzymology. McGraw Hill, New York, p 288
Jencks WP 1975 Binding-energy, specificity, and enzymic catalysis: Circe effect. Adv Enzymol Relat Areas Mol Biol 43:219–410
Jones EW, Fink GR 1982 Amino acid and nucleotide biosynthesis. In: Strathern JN, Jones EW, Broach JR (eds) The molecular biology of the yeast *Saccharomyces*: metabolism and gene expression. Cold Spring Harbor Laboratory, Cold Spring Harbor, NY, p 181–299
Page MI, Jencks WP 1971 Entropic contributions to rate accelerations in enzymic and intramolecular reactions and the chelate effect. Proc Natl Acad Sci USA 68:1678–1683
Raasch MS 1980 Annelations with tetrachlorothiophene 1,1-dioxide. J Org Chem 45:856–867
Sauer J 1966 Diels–Alder reactions: new preparative aspects. Angew Chem Int Ed Engl 5:211–230
Sauer J 1967 Diels–Alder reactions: the reaction mechanism. Angew Chem Int Ed Engl 6:16–33
Sogo SG, Widlanski TS, Hoare JH, Grimshaw CE, Berchtold GA, Knowles JR 1984 Stereochemistry of the rearrangement of chorismate to prephenate: chorismate mutase involves a chair transition state. J Am Chem Soc 106:2701–2703
Weiss U, Edwards JM 1980 The biosynthesis of aromatic amino compounds. Wiley, New York, p 134–184
Ziegler FE 1988 The thermal aliphatic Claisen rearrangement. Chem Rev 88:1423–1452

DISCUSSION

Mountain: Can you say a little more about your mutagenesis strategy?

Hilvert: For the mutagenesis experiments described here we treated whole cells harbouring the chorismate mutase abzyme with ethyl methanesulphonate. This is a crude approach and will produce lesions in both the host genome and the plasmid carrying the antibody genes. Genetic analysis of the mutants will allow us to sort out the details. To restrict mutations to the antibody-encoding genes, we are mutagenizing the plasmid directly with hydroxylamine *in vitro*. We are also using 'window' mutagenesis as described by Hermes et al (1990), to restrict changes to relevant portions of the antibody surface, i.e. the hypervariable loops.

Blackburn: In the engineered yeast cells, how is the second step of the transformation carried out? Chorismate mutase usually not only does the sigmatropic rearrangement but also catalyses the dehydration step.

Hilvert: Yeast chorismate mutase is a monofunctional enzyme: unlike the *E. coli* enzyme it is not fused to either prephenate dehydratase or prephenate dehydrogenase. This is one reason we chose yeast as the host organism for our selection experiments. Keith Backman at BioTechnica International Inc (Cambridge, MA) has genetically separated the *E. coli* chorismate mutase activity from prephenate dehydratase and prephenate dehydrogenase (Maruya et al 1987), so one could now use *E. coli* as a host for the antibody. We are doing those experiments, as is Peter Schultz.

Schultz: You can also do the complementation thermally: if you grow the bacteria at high temperature, the prephenate undergoes a thermal secondary reaction.

Harris: Was the antibody that you expressed in yeast catalytic when you isolated it originally?

Hilvert: Yes. It had the same specific activity as the chorismate mutase antibody isolated from hybridoma cells.

Schwabacher: There is a penalty to the yeast cell in making large amounts of a 'foreign' protein: the cell may not be converting enough chorismate to supply the extra phenylalanine required to make the antibody protein.

Hilvert: Yeast cells producing large amounts of a foreign antibody are not particularly healthy.

Plückthun: Can you estimate what fraction of the immunoglobulin synthesized in the yeast cells actually folds up and assembles? If you did affinity chromatography with any of your transition state analogues as ligands and determined the amount of material that can be retained versus the amount that goes through, then you could estimate the fraction of functional antibodies.

Hilvert: We estimate by ELISA that 60–70% of the light chain produced in yeast folds properly and assembles with heavy chain to create a functional Fab. In addition, the intracellular antibody appears to be relatively long lived in the strain of yeast we use. Once synthesized, the antibody is not rapidly metabolized in the cell.

Mountain: Do you have a feel for what degree of improvement you need in order to get complementation?

Hilvert: My guess is that we need 10–100-fold improvement for that.

Harris: You may be better off putting the expression construct into the yeast chromosome, rather than leaving it on extrachromosomal high copy number plasmids.

Hilvert: That's quite possible. We have isolated one mutant in which a gene conversion event has apparently occurred. The mutant yeast cells produce antibody protein, but we cannot isolate the plasmid from them. I believe the plasmid has been incorporated into the yeast genome. If recombination has occurred, the cells will have only one copy of the antibody-encoding genes which will limit protein production. We are therefore trying to isolate and characterize the antibody from these cells to see whether the apparent complementation is due to increased chemical efficiency of the abzyme.

Harris: In the original experiments, where *E. coli* mutants were complemented by *Neurospora* genes, not much expression was needed to get good complementation of a mutation (Ratzkin & Carbon 1977).

Hilvert: The *leu2d* gene that we are using as a selectable marker was identified that way.

Benkovic: How does expression in yeast compare with that in the bacterial system?

Hilvert: I don't know about Andreas Plückthun's bacterial system, but we have put the same chorismate mutase antibody into *E. coli* and are getting much lower expression there than in yeast.

Plückthun: It is only reasonable to quote a number in $mg\,l^{-1}OD_{550}{}^{-1}$, because one can, of course, grow yeast and *E. coli* to tremendous densities. Typical values of purified antibody fragments are around 0.5–2.0 $mg\,l^{-1}OD_{550}{}^{-1}$.

Hilvert: For practical production, *E. coli* would likely be the vector of choice if expression levels can be improved.

Plückthun: There are probably advantages to both. One advantage of *E. coli* is that the transformation is very efficient, so one can transform very large libraries and get a considerable variety of mutants from plasmid mutagenesis, using the 'dirty oligonucleotide techniques' that e.g. Jeremy Knowles has developed (Hermes et al 1990).

Hilvert: The transformation efficiency of yeast is probably 100-fold less than that in *E. coli*. On the other hand, no one has successfully expressed a functional antibody inside the cytoplasm of *E. coli*.

Plückthun: There may be physical reasons for that; I believe that it has to do with the disulphide bonds. Perhaps there is a difference between the cytoplasm of yeast and *E. coli*. It is very intriguing why functional expression doesn't happen in *E. coli*.

Mountain: Cabilly (1989) has described functional expression of Fab fragments in *E. coli*.

Plückthun: One has to be very careful about what percentage of the oxidized protein is obtained only *after* the cells have been opened. That's a very difficult question to decide.

Jencks: Dr Hilvert, have you tried adding a series of alkyl substituents to your substrate in the Diels–Alder reaction to see what effects that has on the kinetic parameters?

Hilvert: We have shown that the reaction rate at low substrate concentrations increases as the length of the alkyl chain increases. We are now in the process of determining the individual kinetic parameters for these substrates.

Blackburn: Antibodies may catalyse reactions by mechanisms different from those normally used by enzymes and in *in vitro* processes. Paul Bartlett showed a big difference in the inhibitory powers of *endo*- and *exo*-isomers of his inhibitor for natural chorismate mutase, which would agree with a 'chair' transition state. There is no reason why an antibody shouldn't operate by a 'boat' transition state. Bartlett's weaker inhibitor might induce an antibody that operates well, but would work by a boat transition state.

Hilvert: That transition state analogue need not necessarily yield a boat-like transition state. We have immunized with that analogue and have obtained antibodies against it, one of which is catalytic. It was less efficient than the abzyme we have characterized here.

Blackburn: You can enhance the *exo*-geometry by modifying Bartlett's inhibitor, by putting a methyl group adjacent to the carboxyl function.

Hilvert: Paul Bartlett (Bartlett et al 1988) says that the conformation of the *exo*-isomer of the chorismate mutase transition state analogue is not boat like. We have immunized mice with the *exo*-analogue and obtained antibodies against it, one of which is catalytic. The catalyst is less efficient than antibody 1F7, but we can't say anything yet about whether the transition state structure is chair or boat like.

Schwabacher: For antibodies that catalyse the Diels–Alder reaction, if you assume the transition state analogue that you have is a perfect analogue, there are still many ways in which an antibody could recognize it. In screening for antibodies with the most desirable catalytic properties, it may be useful to consider Bill Jencks' description (1981) of binding energy attribution. The binding energy of a molecule can be described as the sum of the binding energies for each of two halves of the molecule, plus a connection energy. Effective catalysis of a Diels–Alder reaction should require not high affinity binding of substrates but a high effective molarity between bound substrates. Consequently, an ELISA that measures binding to the transition state analogue in the presence of both substrates (in an unreactive form) will not detect binding that is specific to one piece; the signal it gives will be directly dependent on the connection energy. This should correspond to the effective molarity between the groups, or the advantage to k_{cat}.

Hilvert: Both substrates have quite high K_m values: for *N*-ethylmaleimide K_m is about 20 mM; the K_m for the tetrachlorothiophene dioxide is probably in that range as well. Consequently, these materials do not compete effectively in ELISA with the transition state analogue, which binds to the antibody many orders of magnitude more tightly.

Schwabacher: Those values are for the catalytic antibody, 1E9. What about binding to the other antibodies you isolated that are less efficient catalysts?

Hilvert: We don't know about that.

Schwabacher: My contention is that the non-catalytic antibodies should bind more tightly than do corresponding catalytic ones with similar affinities for the transition state analogue.

Schultz: What is the rate constant for extrusion of the sulphur dioxide?

Hilvert: We don't know; we have to determine that. The chelotropic elimination reaction is much faster than the cycloaddition thermally, and may also be faster in the presence of the catalytic antibody.

References

Bartlett PA, Nakagawa Y, Johnson CR, Reich SH, Luis A 1988 Chorismate mutase inhibitors: synthesis and evaluation of some potential transition-state analogues. J Am Chem Soc 53:3195–3210

Cabilly S 1989 Growth at sub-optimal temperatures allows the production of functional, antigen-binding Fab fragments in *Escherichia coli*. Gene 85:553–557

Hermes JD, Blacklow SC, Knowles JR 1990 Searching sequence space by definably random mutagenesis: improving the catalytic potency of an enzyme. Proc Natl Acad Sci USA 87:696–700

Jencks WP 1981 On the attribution and additivity of binding energies. Proc Natl Acad Sci USA 78:4046–4050

Maruya A, O'Connor MJ, Backman K 1987 Genetic separability of the chorismate mutase and prephenate dehydrogenase components of the *Escherichia coli tyrA* gene product. J Bacteriol 169:4852–4853

Ratzkin B, Carbon J 1977 Functional expression of cloned yeast DNA in *Escherichia coli*. Proc Natl Acad Sci USA 74:487–491

Characterization of the mechanism of action of a catalytic antibody

Mark T. Martin* Allen R. Schantz*, Peter G. Schultz† and Anthony R. Rees°

*IGEN Inc, 1530 East Jefferson Street, Rockville, MD 20852, USA, †Department of Chemistry, University of California, Berkeley, CA 94720, USA and °Department of Biochemistry, University of Bath, Bath, UK

Abstract. The time course of phenylacetate hydrolysis by the catalytic antibody 20G9 has a kinetic burst lasting several reaction cycles. The burst is caused by partial mixed inhibition by one product of the hydrolysis, phenol, which binds with an apparent dissociation constant of 4.6 μM. Phenol binding causes k_{cat} to decrease from 9.1 min^{-1} to 1.0 min^{-1} and K_m to decrease from 300 μM to 36 μM. Because K_m decreases but k_{cat}/K_m is unaffected, phenol must perturb the ground state structure but not the transition state structure. Structural complementarity to the transition state seems to be an important contributor to catalysis by 20G9 because weak binding in the ground state can be markedly improved by adding phenol, but tight binding of the transition state, which has been optimized by the immune system, cannot be readily improved. Further evidence that the substrate ground and transition states differ greatly in complementarity to the antibody is that the substrate binds more than five orders of magnitude more weakly than the transition state analogue hapten to which the antibody was raised. Two additional phenol molecules bind at higher product concentrations; the first binds over the concentration range of 15 to 86 μM and accelerates hydrolytic activity by 42%; the second is a competitive inhibitor with a K_i of 140 μM. Binding of multiple phenol molecules suggests the presence of abundant hydrophobic amino acids in the complementarity-determining region of 20G9.

1991 Catalytic antibodies. Wiley, Chichester (Ciba Foundation Symposium 159) p 188–200

To develop better catalytic antibodies we need to study the kinetic and structural details of the existing ones. Inhibitors are invaluable in the study of enzyme mechanism, structure and modulation (e.g. Walsh 1984, von Bahr-Lindstrom et al 1981, Martin et al 1987, respectively) and can be expected to become important probes of catalytic antibody structure and function in the future. In this paper we describe a study of the inhibition of a catalytic antibody by the product of its reaction and discuss the mechanistic and structural implications of the results.

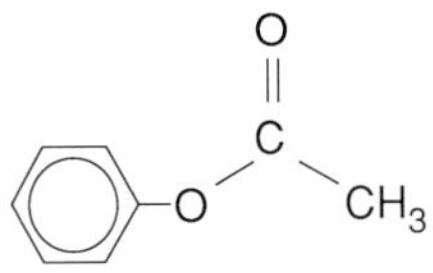

FIG. 1. Structures of phenyl acetate **1** and phenyl phosphonate **2**, a mimic of the postulated transition state in the hydrolysis of phenyl acetate.

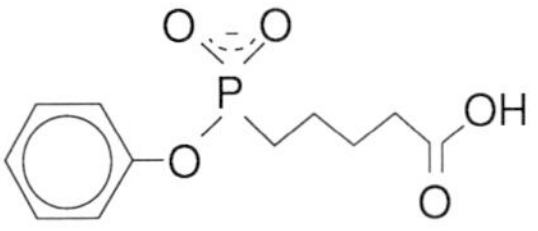

The catalytic monoclonal antibody, 20G9 (Durfor et al 1988), was raised to the phenyl phosphonate hapten, **2**, a mimic of the postulated tetrahedral transition state in the hydrolysis of phenyl acetate **1** (Fig. 1). 20G9 is active in purely aqueous solution and in reverse micelles (Durfor et al 1988) and functions as the molecular recognition element in a biosensor (Blackburn et al 1990).

Modulation of the activity of catalytic antibody 20G9 by phenol

The hydrolysis of phenyl acetate by 20G9 is characterized by a burst detectable at initial substrate concentrations of about 50 μM and above (Fig. 2). The burst is not caused by a rate-limiting catalytic step such as deacylation of a covalently bound intermediate because it lasts for several rounds of hydrolysis. After the reaction has reached its steady-state velocity, addition of more antibody or substrate does not cause a renewed burst, indicating that the burst is not due to slow binding of substrate, a slow structural change in the substrate or antibody on dilution, or acylation by phenyl acetate. Initiation of the assay in the presence of one of the hydrolysis products, acetate (1.0 mM), has no effect on the appearance of the burst (under conditions in which the burst is essentially complete when 10 μM acetate has been formed), but when 12.5 μM phenol is present the burst is essentially eliminated and only the final steady-state velocity is observed (Fig. 2). Thus, activity deceleration is caused by the product, phenol. This is supported by the observation that phenol added at lower concentrations (1–10 μM) only partially eliminates the burst.

To estimate the initial (V_0) and final (V_f) velocities of bursts, we used the equation (1), where P is product, k_{obsd} is an apparent first order rate constant

$$[P] = V_f \cdot t + ((V_0 - V_f)(1 - \exp(-k_{obsd} \cdot t))/k_{obsd} + d \qquad (1)$$

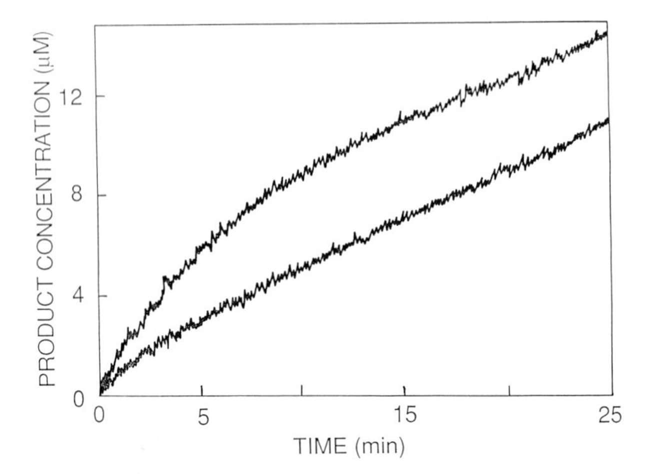

FIG. 2. Progress curves of the hydrolysis of 200 μM phenyl acetate by monoclonal antibody 20G9 (0.79 μM active site concentration) begun in the absence (upper curve) and presence (lower curve) of 12.5 μM phenol. Assays were performed by measuring the absorbance change at 270 nm ($\Delta\epsilon = 767\ M^{-1}cm^{-1}$) at 25 ± 0.1 °C in 10 mM Tris, pH 8.8, containing 140 mM NaCl. Contribution of background hydrolysis was corrected for by including the same solution in a reference cuvette as in the assay cuvette, except antibody.

for formation of the less active antibody species, and d is an offset factor to allow for curves not passing through the origin. This was fitted to curves showing the progress of the reaction over 20–25 minutes and gave an excellent fit. A plot of the initial rate of phenol formation, V_0 ($= dP/dt$) versus the rate of inactivation, k_{obsd} is linear (data not shown), as would be expected if the decline in activity results from binding of product to the catalytic antibody.

Velocities (V_0 or V_f) at various initial substrate concentrations were fitted directly to the Michaelis–Menten equation using a least squares non-linear regression computer program (Duggleby 1981) which gave the values for the kinetic parameters shown in Table 1. Lineweaver–Burk plots (Fig. 3A) illustrate that k_{cat} and K_m decrease dramatically on phenol binding but k_{cat}/K_m (slope) is unaltered. Also shown in Table 1 are values for the parameters of hydrolysis by the Fab′ fragment. Kinetically, the Fab′ fragment behaves similarly to the whole antibody (Fig. 3B). Because 20G9 is a bivalent immunoglobulin G we investigated the possibility that the burst resulted from intramolecular interactions between the two antigen-binding sites. The monovalent Fab′ fragment of 20G9 was prepared by hydrolysis with immobilized pepsin and its kinetic properties were determined. It displays the burst characteristics and, per mole of binding site, k_{cat}/K_m values for both V_0 and V_f do not differ greatly from those for the whole antibody (Table 1).

TABLE 1 Kinetic parameters for hydrolysis of phenyl acetate by catalytic antibody 20G9

	V_0			V_f		
Antibody	k_{cat} min^{-1}	K_m μM	k_{cat}/K_m $min^{-1}M^{-1}$	k_{cat} min^{-1}	K_m μM	k_{cat}/K_m $min^{-1}M^{-1}$
20G9	9.1	300	3.0×10^4	1.0	36	2.8×10^4
20G9 Fab′	ND	ND	9.3×10^3	0.6	41	1.4×10^4

Conditions: pH 8.8, 10 mM Tris, 150 mM NaCl, 25 °C. ND, not determined.

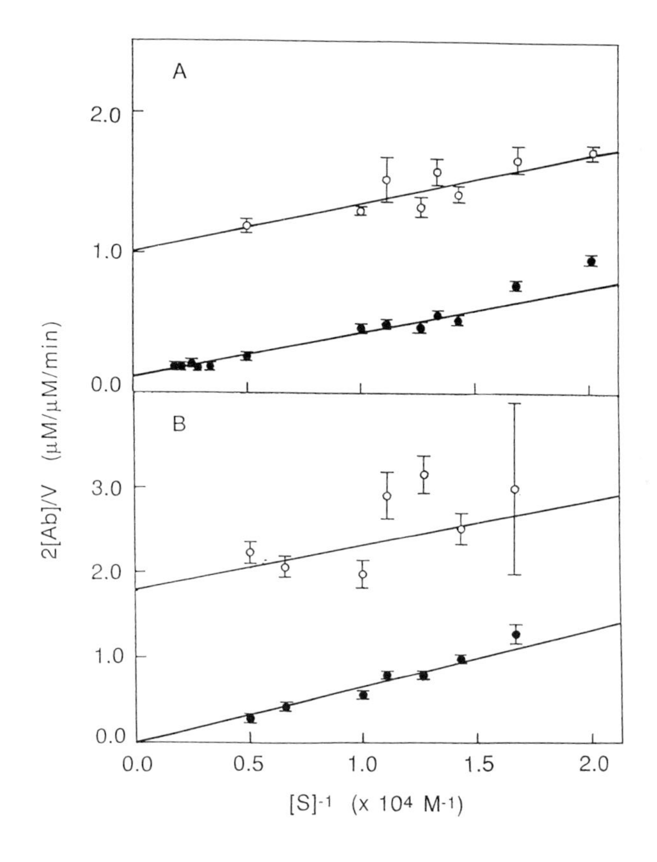

FIG. 3. Double reciprocal plots of initial velocity, V_0 (●), and final steady-state velocity, V_f (○), of phenyl acetate hydrolysis by (A) 20G9 (active site concentration = 0.79 μM), or (B) 20G9 Fab′ fragment (1.23 μM). Isolated 20G9 was converted to $F(ab')_2$ fragments by incubation with insoluble pepsin–agarose, reduction with cysteine, and dialysis. SDS–PAGE showed that the Fab′ had an apparent molecular weight of 46 000.

An apparent dissociation constant for phenol, $K_{p(app)}$, estimated as the product concentration at which $V = V_f + (V_0 - V_f)/2$, was obtained from the reaction curves. By substitution, $K_{p(app)}$ is the product concentration at $t = 0.693/k_{obsd}$. The mean value for $K_{p(app)}$ from eight reaction curves ($S_0 = 50$–$200\ \mu M$) is $4.6 \pm 0.3\ \mu M$.

Non-productive competitive binding of substrate causes identical decreases in both the apparent k_{cat} and K_m values of an enzyme (Bender & Kézdy 1965, Fersht 1974). Applying this model to the hydrolysis reaction, in the absence of phenol there would be a single, productive substrate-binding site on the antibody. Only when phenol was bound would the antibody be able to bind substrate non-productively (in an orientation incorrect for hydrolysis). This newly formed binding mode would overlap the original, productive site, resulting in competitive substrate inhibition. This model is intuitively unattractive, because it requires that phenol binding perturbs the structure of the antibody sufficiently to create a second substrate-binding mode (overlapping the productive site) yet has no effect on the properties of the productive binding mode.

Another model, partial mixed inhibition, depicted in Fig. 4, could also cause identical decreases in both k_{cat} and K_m (Segel 1975). For the hydrolysis of phenyl acetate catalysed by monoclonal antibody 20G9, $k_{cat} = 9.1\ min^{-1}$, $K_m = 300\ \mu M$, $K_{p(app)} = 4.6\ \mu M$, $\alpha = 0.12$ and $\beta = 0.11$. With this model, there is only a single substrate-binding mode. Binding of phenol to the antibody promotes substrate binding decreases (K_m) and decreases the turnover number (k_{cat}), to identical, finite degrees. This model agrees satisfactorily with our observations.

Structural and mechanistic implications of modulation of catalysis by phenol

It is easy to speculate on how the binding of phenol to 20G9 lowers the K_m of hydrolysis of phenyl acetate. Antigen-binding sites have been reported to be rich

$$
\begin{array}{ccccc}
Ab + S & \underset{}{\overset{K_M}{\rightleftharpoons}} & Ab{\cdot}S & \overset{k_{cat}}{\longrightarrow} & Ab + P_1 + P_2 \\
+ & & + & & \\
P_2 & & P_2 & & \\
K_{papp}/\alpha \ \updownarrow & & \updownarrow \ K_{papp} & & \\
Ab{\cdot}P_2 + S & \underset{\alpha K_M}{\rightleftharpoons} & Ab{\cdot}S{\cdot}P_2 & \underset{\beta k_{cat}}{\longrightarrow} & Ab{\cdot}P_2 + P_1 + P_2
\end{array}
$$

FIG. 4. Model of partial mixed inhibition of the hydrolysis of phenyl acetate catalysed by an antibody. Ab, antibody; P_1, acetate; P_2, phenol; S, phenyl acetate.

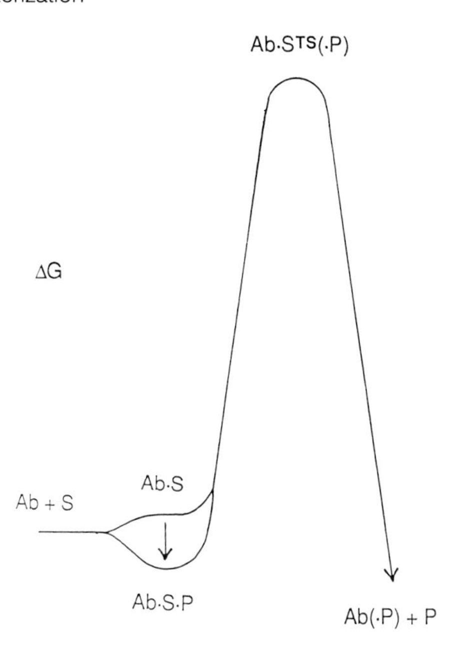

FIG. 5. Gibbs free energy diagram for hydrolysis of phenyl acetate by monoclonal antibody 20G9 in the absence and presence of bound phenol. The diagram shows the free energy profile for 200 μM phenyl acetate (the substrate concentration in the burst shown in Fig. 1). The length of the arrow on the left represents 4.2 kJ/mole.

in aromatic tyrosine and tryptophan residues (Davies et al 1990); thus one possibility is that phenol takes part in hydrophobic or hydrogen-bonding interactions with one or more antibody side chains, causing local conformational changes in the antibody which result in tighter substrate binding. Alternatively, phenol could interact directly with both the antibody and the phenyl ring of the substrate in a sandwich-like hydrophobic ternary complex. Either possibility could result in improved substrate binding in the ground state.

Binding of phenol to 20G9 drastically decreases the apparent ground state dissociation constant, K_m, which in Fig. 5 is reflected by the Gibbs free energy difference (ΔG) between the ground state antibody–substrate complex (Ab·S), and free substrate and antibody (Ab + S). On the other hand, binding of phenol has essentially no effect on the second order rate constant, k_{cat}/K_m, which is related to the free energy difference between the transition state complex (Ab·S^{TS}) and the free substrate and antibody (Fig. 5). Thus, the binding of phenol to the antibody perturbs the structure of the ground state complex but not the structure of the transition state complex (Jencks 1987). In Fig. 5, k_{cat} relates to the difference in free energy between the transition state (Ab·S^{TS}) and the ground state (Ab·S). Binding of product, P, simultaneously decreases K_m and k_{cat}.

In the absence of phenol the substrate binds to the antibody weakly with a K_m of 300 μM, which is at the upper end of the range of antigen–antibody dissociation constants (Schultz 1989). Weak ground state binding here is understandable, because the antibody is structurally complementary, not to phenyl acetate, but to the phenyl phosphonate hapten. The two structures differ substantially, especially in the vicinity of the scissile bond (Fig. 1). Because the substrate is weakly bound by the antibody, the complex should be easily perturbed by changes in the antibody binding site, such as those caused by binding of phenol. When phenol binds to the antibody it increases antibody–substrate affinity, stabilizing the ground state complex, and the K_m for phenyl acetate is greatly reduced.

On the other hand, phenol does not affect transition state interactions. The immune system has created a binding site structure complementary to the tetrahedral phosphonate hapten and, in theory, the transition state of the hydrolysis reaction. If the antibody functions by being structurally complementary to the reactant transition state, there will be a large number of favourable binding interactions exclusive to the transition state. It is therefore less likely that phenol affects the optimized binding interactions between the antibody and the transition state.

Another piece of evidence that 20G9 uses transition state structural complementarity to hydrolyse phenyl acetate is that the substrate (in the absence of product) binds quite weakly to 20G9 with a K_m of 300 μM, whereas the transition state analogue (hapten) binds very tightly with a K_i of 2.2 nM (Blackburn et al 1990). The difference between the apparent dissociation constants must result from specific structural differences between the hapten–antibody and substrate–antibody complexes. Much of the hapten–antibody complementarity is localized in the hapten tetrahedral phosphonate group; the remainder of the substrate and hapten structures are similar (Fig. 1). The hapten also has a free carboxylate, missing in the substrate, that could contribute binding energy. However, this carboxylate was blocked with a carrier protein during immunization, therefore any charged binding interactions between it and the antibody are non-specific and coincidental.

If 20G9 does function solely by being structurally complementary to the transition state, and the hapten is a perfect mimic of the transition state, then K_{TS} will equal K_i, where K_{TS} is the dissociation constant of the hydrolytic transition state (Wolfenden 1976). The expected value of K_i (K_{TS}) may be estimated experimentally by $K_{TS} = K_m \times k_{uncat}/k_{cat}$, where k_{uncat} is the pseudo first order uncatalysed rate of hydrolysis of phenyl acetate by hydroxide ion (under our conditions $k_{uncat} = 6.1 \times 10^{-4}\ \text{min}^{-1}$). This concept has been used previously as evidence for complementarity of an antibody to a transition state (Benkovic et al 1988, Janda et al 1989). If other contributing factors are involved, such as nucleophilic catalysis, the value of K_{TS} will be less than that of K_i (Benkovic et al 1988). For 20G9, the value of K_{TS} is 20.1 nM, ninefold greater

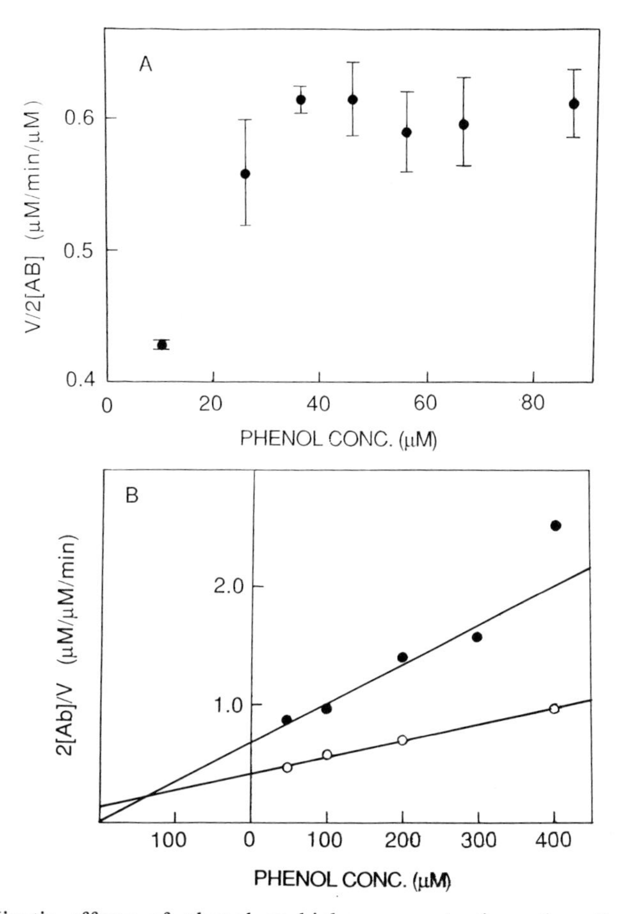

FIG. 6. Kinetic effects of phenol at higher concentrations than those typically encountered during an assay. (A) Effect of increasing concentrations of phenol on the steady-state velocity of a reaction. Initial phenyl acetate concentration was 200 μM. (B) Dixon plot of inhibition of monoclonal antibody 20G9 by concentrations of phenol higher than normally encountered in an assay. Inhibition was measured at initial phenyl acetate concentrations of 100 μM (●) and 200 μM (○).

than the K_i of 2.2 nM; in other words, binding of the hapten releases 5.46 kJ/mol more Gibbs free energy than would be expected. This value is not unreasonable, given the structural differences between the hapten and the substrate; the main difference is that the hapten has a free carboxyl group that may provide binding interactions with the antibody. Also, it is unlikely that the phosphonate structure is an ideal mimic of the transition state. What can be concluded from a comparison of K_{TS} and K_i is that factors in addition to transition state complementarity (general acid-base contributions, for instance) need not be invoked to explain catalysis by 20G9.

The effect of high concentrations of phenol on the activity of catalytic antibody 20G9

When phenol was added at concentrations greater than those typically produced during an assay, additional kinetic effects were observed. At an initial substrate concentration of 200 μM, increasing the concentration of added phenol from 11 μM (that formed by the antibody at steady state in the absence of added phenol) to 86 μM caused a 42% increase in the steady-state velocity (Fig. 6A). No burst was observed in this phenol concentration range. At phenol concentrations higher than approximately 86 μM, the catalytic activity of 20G9 declines. This decrease appears to be due to competitive inhibition by phenol. The K_i for this inhibition was found to be 140 μM from the Dixon plot (Dixon 1953) shown in Fig. 6B.

It thus appears that there are three phenol-binding modes of 20G9 capable of altering the kinetic parameters of phenyl acetate hydrolysis: one causes the burst, a second causes partial reactivation, and a third causes competitive inhibition. This may reflect a large number of hydrophobic residues in or very near the active site of the antibody. If hydrophobic binding sites are typically present in antigen-binding sites, small, hydrophobic substrates or products may in general alter the kinetics of reactions catalysed by antibodies.

Acknowledgements

We thank Drs Richard J. Massey and Andrew Napper for constructive criticism during preparation of the manuscript and Dr Renee Sugasawara for preparation of the antibody.

References

Bender ML, Kézdy FJ 1965 Mechanism of action of proteolytic enzymes. Annu Rev Biochem 34:49–76

Benkovic SJ, Napper AD, Lerner RA 1988 Catalysis of a stereospecific bimolecular amide synthesis by an antibody. Proc Natl Acad Sci USA 85:5355–5358

Blackburn GF, Talley DB, Booth PM et al 1990 Potentiometric biosensor employing catalytic antibodies as the molecular recognition element. Anal Chem 62:2211–2216

Davies DR, Padlan EA, Sheriff S 1990 Antibody–antigen complexes. Annu Rev Biochem 59:439–473

Dixon M 1953 The determination of enzyme inhibitor constants. J Biochem 55:170–171

Duggleby RG 1981 A nonlinear regression program for small computers. Anal Biochem 110:9–18

Durfor CN, Bolin RJ, Sugasawara RJ, Massey RJ, Jacobs JW, Schultz PG 1988 Antibody catalysis in reverse micelles. J Am Chem Soc 110:8713–8714

Fersht AR 1974 Catalysis, binding and enzyme–substrate complementarity. Proc R Soc Lond B Biol Sci 187:397–407

Janda KD, Benkovic SJ, Lerner RA 1989 Catalytic antibodies with lipase activity and R or S substrate specificity. Science (Wash DC) 244:437–440

Jencks WP 1987 Economics of enzyme catalysis. Cold Spring Harbor Symp Quant Biol 52:65–73

Martin M, Vallee BL, Riordan JF 1987 Fluorescent inhibitor probes of enzyme active site conformation: anion binding to angiotensin converting enzyme. Anal Biochem 161:341–347
Schultz PG 1989 Catalytic antibodies. Acc Chem Res 22:287–294
Segel IH 1975 Enzyme kinetics. John Wiley, New York, p 188
von Bahr-Lindstrom H, Sohn S, Woenckhaus C, Jenk R, Jörnvall H 1981 Characterization of a structure close to the coenzyme-binding site of liver aldehyde dehydrogenase. Eur J Biochem 117:521–526
Walsh CT 1984 Suicide substrates, mechanism-based enzyme inactivators: recent developments. Annu Rev Biochem 53:493–535
Wolfenden R 1976 Transition state analog inhibitors and enzyme catalysis. Annu Rev Biophys Bioeng 5:271–306

DISCUSSION

Arata: Do you have any direct experimental evidence that phenol molecules are bound to the antigen-binding site?

Martin: All our evidence is from kinetics, no direct physical experiments were carried out.

Arata: This could be done by NMR.

Leatherbarrow: You fitted your data to a slow-binding equation, but you said that the reaction curves showed product inhibition. Are the two related? I don't see that one equation follows from the other.

Martin: In theory they should be related, provided that in the region of the burst being fitted to the equation there is little substrate depletion and the phenol concentration is significantly greater than the antibody concentration.

Leatherbarrow: Slow-binding curves comprise the sum of exponential and linear components. The absorbance, A, varies with time, t, as:

$$A = v_s t + (v_0 - v_s)(1 - e^{-kt})/k$$

where v_0 and v_s are the initial and final rates and k is the rate constant for the binding. Product inhibition involves a completely different (and far more complex) equation (for example, see Cornish-Bowden 1979). The curves produced by each equation may look similar but they involve different constants and a different mathematical form.

Martin: The experimentally obtained curves and the theoretical fits are almost identical; if there is a difference, it is insignificant.

Leatherbarrow: I am sure the data fit; I am questioning whether the slow-binding equation is appropriate. You are saying that the mechanism is product inhibition, but you are fitting a slow-binding equation. The two are not the same.

Benkovic: You could tell whether it's slow binding by other experiments, as Dr Martin has done, by adding phenol and then adding the substrate and monitoring the onset of inhibition as a function of time.

Jencks: Is radioactively labelled phenol incorporated into the substrate?

Martin: We didn't use radioactive material. The only experiment that we've done along those lines is to have relatively high concentrations of aniline present while the antibody hydrolysed phenyl acetate. After a time, we separated the mixture by HPLC under conditions where we could detect very low concentrations of acetanilide: none was found.

Jencks: Have you looked at substituted phenyl acetates as substrates?

Martin: That was done by Gary Blackburn at IGEN (Blackburn et al 1990). He used this antibody as the molecular recognition element in a potentiometric biosensor. The biosensor consisted of essentially electrometers that measured the potential of a modified pH-sensitive electrode relative to reference electrodes. The tip of the modified pH electrode was covered by a dialysis membrane that entrapped the catalytic antibody. When phenyl acetate or another substrate came into contact with the end of the electrode, it was hydrolysed and the liberated protons were detected by the pH electrode.

Of 13 structurally related compounds tested, phenyl acetate was the best substrate (it gave the strongest response in the biosensor), followed by compounds that had substitutions to the methyl group of phenyl acetate in the same position as the linker through which the hapten was coupled to the carrier protein during immunization. Poorer substrates had nitro and methyl substituents on the phenyl ring. Acetanilide, methyl benzoate and acetylcholine were not detectably hydrolysed.

Janda: Can you tell us a little more about the biosensor? Is it a linear response? How stable is it in aqueous and organic solvents?

Martin: Biosensors are increasingly being used to detect low concentrations of analytes. There are two main types of biosensors, those that use catalysts (enzymes) to recognize the analyte and those that use molecules that only bind the analyte, such as antibodies. Enzyme-based catalytic biosensors have the advantages of being reusable and very efficient, but the drawback of being limited to reactions that occur in Nature. Also, a chosen enzyme often cannot be isolated in a stable form or in large quantities.

Biosensors that work by binding of monoclonal antibodies have the advantage that they can specifically recognize an enormous range of compounds that enzymes cannot. Antibodies are stable and can be produced in large amounts by hybridoma technology. They do have the problems that they are not easily reusable—once an analyte is bound, it stays bound, and the mere act of binding is rarely accompanied by physical changes large enough to be transduced into a measurable signal. Obviously, catalytic antibodies are a great idea for biosensors, because they combine the best features of antibodies and enzymes. This biosensor was simply a prototype, since there is not much of a market for measuring low concentrations of phenyl acetate. It showed substrate specificity, was inhibited by hapten (one can also use biosensors to detect inhibitors), and had a lower detection limit of 5 μM phenyl acetate. After seven months or so, the biosensor still functioned, so it was very stable. It has only been looked at in aqueous solution.

Page: Do you see the catalytic burst with all the esters?

Martin: No, because this was done on the biosensor. The antibody concentration in the electrode of the biosensor was 50–100 mg/ml (0.7–1.3 mM), so it was not Michaelis–Menten kinetics.

Plückthun: I presume that when you say that the specific catalytic activity of the Fab fragment is only half that of the whole antibody, you have taken into account the molarity of the binding sites?

Martin: We routinely determine the antibody concentration by active site titration with the hapten inhibitor, both for the whole antibody molecule and for the Fab fragment. So if the two- to threefold decrease in k_{cat}/K_m seen with the Fab fragment *is* due to damage to the protein during preparation, I am forced to assume that the Fab has retained the ability to bind the hapten tightly but has lost some of its catalytic activity. I can't rule out the possibility that the Fab preparation is perfectly healthy but simply has different kinetic properties to the whole IgG.

Page: Did you measure the pH dependence of the hydrolysis?

Martin: Yes. That was done in the biosensor work. An increase in biosensor response, or catalytic activity, was seen over the range pH 7.3–9.4.

Harris: Is the hapten presented from the outside of the presenting cell or from the inside of the cell after processing? The method of presentation may affect which parts of the hapten are 'seen'. I ask this because of Don Hilvert's hierarchy of reactivity with ethyl, methyl and isopropyl substitutents. That alkyl group was the attachment site to the KLH for your hapten, wasn't it?

Hilvert: Yes. We attached the hexachloronorbornene transition state analogue to KLH via a hexanoic acid. The resulting antibodies exhibit substrate specificities that reflect the length of the *N*-alkyl substituent on maleimide.

Kang: Recognition of linkers by antibodies was a problem in developing conventional immunoassays, but has been less of a problem since the introduction of heterologous linkers.

Plückthun: The importance of the carrier and the linker seems to vary from system to system. There's an extensive literature on phosphorylcholine as a hapten, which has been used to derivatize proteins and is the natural surface antigen of several bacteria, where it is bound to a polysaccharide. The binding constants and cross-reactivities don't seem to depend on what you immunize with. With the phosphorylcholine hapten, the antibodies use a pre-formed cavity that is encoded in the germline. This class of anti-phosphorylcholine antibodies seems to completely ignore the linker; but this may be a special case.

Hansen: In practical terms, the question is how much of the linker one should include in the substrate. It seems that sometimes it doesn't matter, and you can include very little of the linker; at other times, it's critically important to extend the substrate to the entire linker.

Lerner: This is the whole issue of whether you want the leaving group to be involved in the immunogenicity or not. We have thought about it both ways. Does one want the antigen to be congruent with the leaving group or does one want it to be totally different? For example, what happens if Mark Martin tries *p*-nitrophenyl acetate?

Martin: p-Nitrophenyl acetate is hydrolysed roughly one-fifth as fast as phenyl acetate in the biosensor. In our case, changing the leaving group slows the reaction, most likely by hindering initial binding of the substrate. We haven't investigated it, but what we lose in poorer substrate binding may be regained over the time of the reaction in diminished product inhibition, as Dr Benkovic described.

As a detection method, we have added fluorophores to the ends of substrates. It actually enhances binding to have, for example, a dansyl group essentially taking up space where the linker and carrier protein were during immunization. There is no *a priori* reason to do it, but adding things to the carrier protein end of the substrate may in general help binding.

Shokat: Perhaps the effect of the dansyl group is the same as that of the alkyl group in Don Hilvert's dienophile—just a 'salting in' effect driving the substrate into the binding site.

References

Blackburn GF, Talley DB, Booth PM et al 1990 Potentiometric biosensor employing catalytic antibodies as the molecular recognition element. Anal Chem 62:2211–2216

Cornish-Bowden A 1979 Fundamentals of enzyme kinetics. Butterworth, London

Catalytic antibodies: a new window on protein chemistry

Colin J. Suckling*, Catriona M. Tedford*, George R. Proctor*, Abedawn I. Khalaf*, Laura M. Bence† and William H. Stimson†

Departments of Pure and Applied Chemistry and Bioscience and Biotechnology†, University of Strathclyde, 295 Cathedral Street, Glasgow G1 1XL, UK*

Abstract. Catalytic antibodies have created a new dimension in protein chemistry. In these studies it is particularly valuable to investigate systems for which natural enzymic catalysts are unknown. At Strathclyde we have examined several ways of preparing homochiral building blocks for organic synthesis. Antibodies that catalyse the Diels–Alder reaction have been characterized. The target reaction was the addition of acetoxybutadiene to *N*-substituted maleimides, a reaction that should give a pentafunctional homochiral building block. Catalytic antibodies can give insight into the mechanism of catalysis by proteins. We have investigated an adventitious hydrolytic antibody that cleaves activated esters. We have also shown that an antibody raised to ampicillin for analytical purposes catalyses hydrolysis of the β-lactam ring.

1991 Catalytic antibodies. Wiley, Chichester (Ciba Foundation Symposium 159) p 201–210

Few innovations in the field of biological chemistry can have generated more obvious and immediate opportunities than the discovery of catalytic antibodies. Not only has a new window been opened on the science of protein chemistry and catalysis, but commercial application is a viable proposition. At Strathclyde, we were fortunate to bring to the subject of catalytic antibodies our experience in enzyme inhibition, applications of enzymes to synthesis, and immunology. The central stimulus for our work has been the investigation of reactions of potential interest in synthetic transformations. Clearly, such reactions would best be selected from those for which there are no known enyzmic catalysts. It is important to obtain high selectivity or high rate enhancements, and preferably both. A catalytic antibody will, however, be of interest in synthesis if it catalyses the production of an enantiomerically pure building block from which complex optically pure compounds can be synthesized. The importance of this concept is well illustrated by recent developments in the chemistry of benzene-*cis*-glycol (Widdowson & Ribbons 1990).

Abzymes and the Diels–Alder reaction

Selection of a suitable reaction

The importance of the Diels–Alder reaction in providing building blocks of well-defined stereochemistry for organic synthesis has been recognized and exploited continuously since Woodward's syntheses of cholesterol and cortisone (Woodward et al 1952). The requirements today are more stringent in that absolute configuration as well as relative configuration must be considered in synthetic reactions. The control of chirality has been the most powerful driving force in the development of the application of enzymes to organic synthesis (Suckling 1990). The discovery that antibodies have the ability to catalyse acyl transfer reactions with high stereoselectivity (Napper et al 1987, Janda et al 1989) encouraged us to develop antibodies for chiral Diels–Alder cycloadditions.

We selected a system that would allow a range of properties of antibodies to be investigated and would provide a potentially useful homochiral building block (Fig. 1, **3**). A reactive diene, acetoxybutadiene, was chosen, partnered by a good dieneophile, an *N*-alkylmaleimide. This choice was very similar to that made by Hilvert et al (1989) in that a suitable hapten could easily be prepared with a linker to the protein attached to the maleimide nitrogen remote from the principal site of reaction. Such a hapten is product like; it has been argued that the transition state for a Diels–Alder reaction resembles the product more closely than it does the starting material (Brown & Houk 1984). However, in contrast to Hilvert, we made no special provision to circumvent possible product inhibition and used a hapten derived directly from the starting materials (**2**, $R=(CH_2)_3CO—$). Whilst this was undoubtedly a risk, our experience in generating monoclonal antibodies for analytical purposes suggested that quite small structural changes in an antigen could lead to enormous changes in affinity for a specific antibody (W. H. Stimson et al, unpublished observations). Because both the acetoxy and *N*-alkyl substituents could be easily varied, we viewed the chances of finding catalytic antibodies to a suitable pair of substrates as acceptable. The target molecule itself (**3**, $R=Et$ or CH_2Ph, $X=OAc$) is most attractive for synthesis and contains many exploitable functionalities: an alkene, a secondary allylic alcohol, two potentially differentiable carbonyl groups, and a nitrogen atom.

Antibody production and properties

The N-(4-carboxybutyl) analogue (**3**) of the product was coupled to bovine serum albumin and used to immunize NZB/BALB/c F1 hybrid female mice (8–12 weeks). This mouse strain has been found in our laboratories to be particularly favourable for the production of high affinity antibodies for analytical applications. Antibodies that bound the hapten were identified using a conventional ELISA. We have also generated antibodies to this hapten by *in vitro*

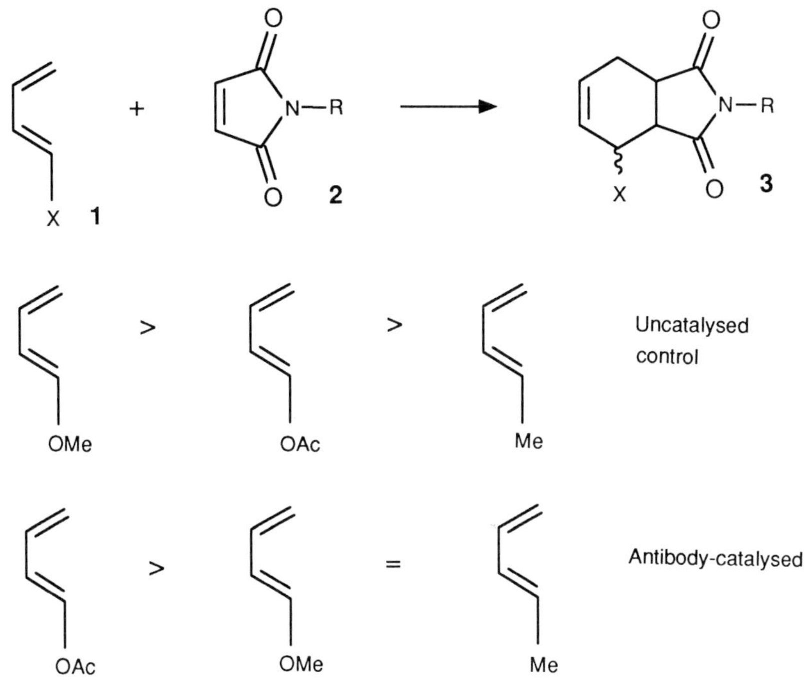

FIG. 1. Diels–Alder reaction catalysis by antibody H11: hapten structure and reactants studied.

immunization techniques but these antibodies have not yet been used for chemical studies. Whilst monoclonal antibodies from immunized mice were being prepared, we investigated briefly the properties of the polyclonal sera. Using high performance liquid chromatography (HPLC) we observed an increase of approximately 100-fold over control in the rate of the reaction and some selectivity with respect to reactants. The most favoured diene was acetoxybutadiene, which was built into the hapten. In comparison, the chemically more reactive methoxybutadiene was less reactive by a factor of at least ten in the presence of the mixture of antibodies. The reactivity of methoxybutadiene was indistinguishable from that of pent-1,3-diene in these preliminary tests. On the other hand, the polyclonal sera showed no preference for reaction with *N*-ethyl- or *N*-benzylmaleimide.

The first monoclonal antibody (designated H11) to be studied in detail bound strongly to the hapten conjugated to transferrin ($K_d = 3.99 \times 10^{-8}$ M) and catalysed the Diels–Alder reaction between acetoxybutadiene and *N*-ethyl- and *N*-benzylmaleimide. Typical reactant concentrations were 4.7×10^{-4} M substrates and 4.0×10^{-5} M antibody H11. The high antibody concentration is consistent with the high K_m observed (8.3 mM). The difference between the values of K_m

and K_d supports the argument made above concerning the tightness of molecular recognition required by the antibody to cause catalysis. No product inhibition was observed at product concentrations up to 1.6 mM. Whilst this observation is scientifically encouraging, millimolar concentrations are still low for preparative reactions of synthetic starting materials. Much will need to be done to obtain a catalyst of practical use.

The rate constant for the antibody-catalysed reaction was $0.055\ s^{-1}$ but we have had some difficulty in determining the bimolecular rate constant for the uncatalysed reaction. The best estimate we have so far for the effective molarity of this reaction is approximately 1 M. This quantitative behaviour agrees with the properties of the mixture of polyclonal antibodies described above. However, with antibody H11 as catalyst no reaction was observed with penta-1,3-diene, and methoxybutadiene showed no rate enhancement over control reactions. H11 shows distinct selectivity for the diene component but some tolerance for the *N*-alkyl group of the dieneophile. Once again, the observation of catalysis and selectivity is encouraging but the low value of k_{cat}/K_m, at least with *N*-ethylmaleimide as substrate, indicates that the performance of this antibody is far from optimal.

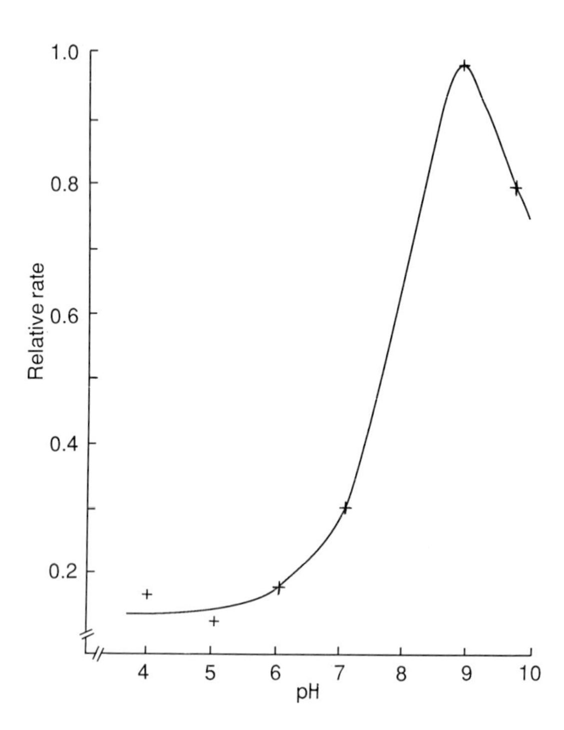

FIG. 2. pH dependence of the reaction of acetoxybutadiene with *N*-ethylmaleimide catalysed by antibody H11.

The reaction of acetoxybutadiene with *N*-ethylmaleimide was pH dependent (Fig. 2), showing a maximum at pH 9. This observation is interesting in that the Diels–Alder reaction is not considered to be pH dependent, although Lewis acid catalysis is well known. The pH dependency could be explained by a conformational change in the antibody. Alternatively, a protonated group such as an imidazolium ring of a histidine residue might interact with the acetoxybutadiene at low pH; this interaction would decrease the electron availability in the diene and hence decrease the reaction rate. More detailed studies are required to establish the role of functional groups in catalysis of this electrocyclic reaction. Lastly, although we have isolated small quantities of product from the Diels–Alder reactions, we have still to determine their chirality. Many further clones of antibodies that bind the hapten remain to be investigated.

Hydrolytic antibodies

Much of the work on catalytic antibodies so far has concerned acyl transfer reactions, especially hydrolyses. Whilst we did not set out to study hydrolytic antibodies themselves, we have discovered some hydrolytic antibodies that exhibit interesting properties. The activity was found simply by screening for transformations of suitable substrates. In the light of the recent recommendations for antibody production by molecular biological methods using the λ vector (Huse et al 1989), it is interesting that simple screening for catalytic activity has already turned up a new abzyme.

Aralkyl ester hydrolysis

The emphasis in hapten design for the generation of catalytic antibodies has been on transition state analogues. Some consideration has been given to the induction of catalytic functional groups in the protein through the inclusion of charged groups within the hapten (Shokat et al 1989, Janda et al 1990). We prepared an aromatic ammonium hapten and raised a number of antibodies to it using the same protocols as outlined for the Diels–Alder reaction. Unfortunately, contractual confidentiality makes it impossible to describe details of these experiments and the full range of properties of the antibodies. One antibody (C3) was found to hydrolyse the 4-nitrophenyl ester of 5-(3-methoxy-phenyl)-pentanoic acid (Fig. 3); 4-chlorophenyl, ethyl and thioethyl esters were not hydrolysed, as shown by HPLC. The rate enhancement was at least 10^6-fold compared to the control at pH 8.5 ($k_{cat} = 2.4\,s^{-1}$, $k_{uncat} = 6.9 \times 10^{-7}\,s^{-1}$). In parallel with the results obtained for the antibody H11 that catalysed the Diels–Alder reaction, the K_m observed (0.4 mM) was much higher than the K_d for the hapten conjugate. The pH rate profile for the hydrolysis reaction showed a plateau above pH 8 (Fig. 4); the shape of the profile suggests that the

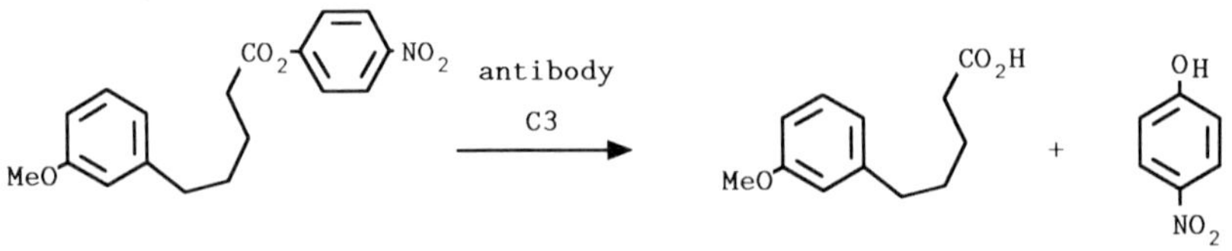

CO_2 NO_2 antibody C3 MeO → MeO CO_2H + OH NO_2

FIG. 3. Hydrolysis of the 4-nitrophenyl ester of 5-(3-methoxyphenyl)-pentanoic acid catalysed by antibody C3.

deprotonation of an imidazolium or carboxylate group may be important in the hydrolysis mechanism. Further details of this mechanism remain to be established. We investigated the possibility that antibody C3 catalysed reactions other than activated ester hydrolysis but neither elimination nor deuterium exchange reactions were observed.

A serendipitous abzyme

The main emphasis in immunology at Strathclyde over recent years has been the production of highly specific, high affinity monoclonal antibodies for analytical purposes. Consequently, a number of antibodies are available for us to evaluate for catalytic properties; many of these are polyfunctional molecules with potential interest for organic chemists. So far, we have found

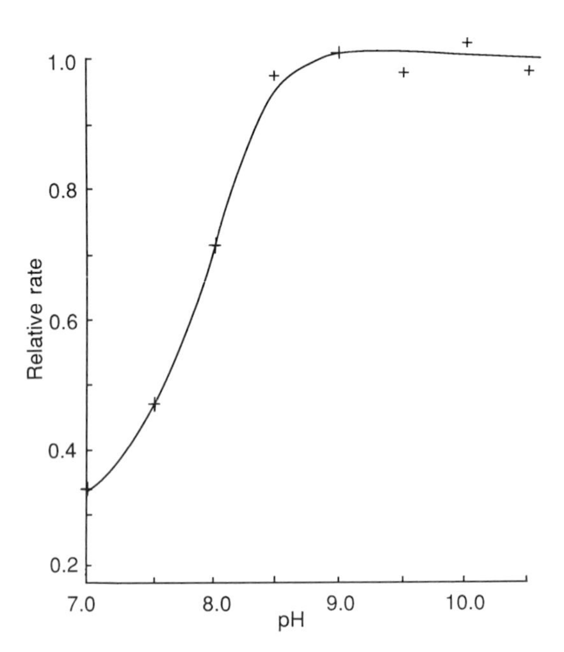

FIG. 4. pH dependence of the hydrolysis catalysed by antibody C3 shown in Fig. 3.

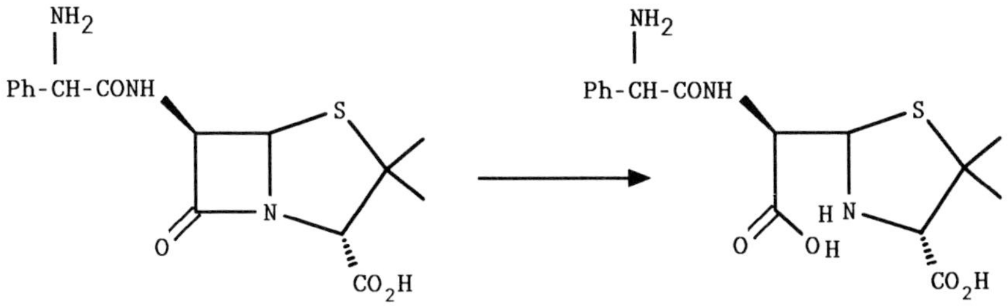

FIG. 5. Hydrolysis of ampicillin by antibodies 2H10 and ID2.

two antibodies with significant properties; 2H10 is a high affinity antibody raised to ampicillin (Fig. 5), ID2 is a lower affinity antibody that binds a wide range of β-lactam antibiotics. Both proteins were found to be β-lactamases. Incubation of each antibody with ampicillin followed by gel permeation chromatography and preparative HPLC of the low molecular weight products identified clearly a compound with the β-lactam ring cleaved but with the side chain amide link intact. The kinetic properties of these antibodies have not yet been investigated.

If our discovery here is not atypical, there must be many catalytic antibodies awaiting discovery in the world's immunology laboratories. The problem, as with antibody libraries generated by molecular biology, will be to find suitably rapid and discriminating screens to identify interesting proteins.

Acknowledgement

We thank the Wolfson Foundation for financial support.

References

Brown FK, Houk KN 1984 The STO-3G transition state structure of the Diels–Alder reaction. Tetrahedron Lett 25:4609–4612

Hilvert D, Hill KW, Nared KN, Auditor M-TM 1989 Antibody catalysis of a Diels–Alder reaction. J Am Chem Soc 111:9261–9262

Huse WD, Sastry L, Iverson SA et al 1989 Generation of a large combinatorial library of the immunoglobulin repertoire in phage lambda. Science (Wash DC) 246:1275–1281

Janda KD, Benkovic SJ, Lerner RA 1989 Catalytic antibodies with lipase activity and R or S substrate selectivity. Science (Wash DC) 244:437–440

Janda KD, Weinhouse MI, Schloeder DM, Lerner RA, Benkovic SJ 1990 Bait and switch strategy for obtaining catalytic antibodies with acyl-transfer capabilities. J Am Chem Soc 112:1274–1275

Napper AD, Benkovic SJ, Tramontano A, Lerner RA 1987 A stereospecific cyclization catalyzed by an antibody. Science (Wash DC) 237:1041–1043

Shokat KM, Leumann CJ, Sugasawara R, Schultz PG 1989 A new strategy for the generation of catalytic antibodies. Nature (Lond) 338:269–271

Suckling CJ 1990 Selectivity in synthesis, chemicals or enzymes. In: Suckling CJ (ed) Enzyme chemistry, impact and applications, 2nd edn. Chapman & Hall, London, p 95–170

Widdowson DA, Ribbons DW 1990 Stereo- and regio-specific enzymic reactions: chemical exploitation of the aromatic dihydroxylation system. In: Copping LG, Martin RE, Pickett JA, Bucke C, Bunch AW (eds) Opportunities in biotransformations. Elsevier Applied Science, London, p 119–130

Woodward RB, Sondheimer F, Taub D, Heusler K, McLamore WM 1952 The total synthesis of steroids. J Am Chem Soc 74:4223–4251

DISCUSSION

Plückthun: Most of the anti-β-lactam antibodies that have been characterized in detail are really directed against the acylated carrier protein, that is, probably the ring-opened form of the β-lactam, or even subsequent rearrangement products. Is it known what antibody 2H10 really recognizes? What was used as the immunogen?

Suckling: These antibodies were prepared by Professor Stimson's colleagues in our Immunology Department. The antibody to ampicillin, 2H10, was raised to a conjugate in which the protein was coupled to the amino group of the side chain of ampicillin. Professor Stimson believes that the β-lactam ring was still intact.

Page: Current opinion seems to be that the penicilloyl amide is the antigen and this is generated by protein lysine residues opening the β-lactam ring.

Schultz: We attempted to raise antibodies to β-lactam antibiotics. We had problems because of hydrolysis of the lactam ring by serum components. It is always a good idea to assay haptens for stability in serum before immunization.

Page: Do you know any details of the rates of that reaction?

Suckling: No, not yet. There was no background reaction under those conditions.

Page: Did you add EDTA? Metal ions can catalyse the hydrolysis of β-lactam antibiotics.

Suckling: We have not done that experiment.

Janda: When you look at other substrates, it would be interesting to look at the saturated analogue of nitrophenol (i.e. 4-nitrocyclohexanol), to see if that is a substrate.

Suckling: There are many obvious experiments of that sort to be done: different structures and leaving groups to be looked at.

Green: Was product released from the antibody? Were there multiple turnovers?

Suckling: We see no inhibition over the initial rate period up to about 50–60%, which is as far as those have been followed. We have isolated product but we haven't characterized its stereochemistry.

Hansen: Which combinations of functionalities at positions X and R in the substrate (Fig. 1) gave the highest activity?

Suckling: Ethyl and benzyl for R are very similar; the only group at X that is recognized by antibody H11 is the acetoxy.

Hilvert: It is very surprising that the *N*-ethyl acetoxy product doesn't inhibit the reaction.

Suckling: I agree, but we have never seen any significant inhibition over the range of concentrations that we have been able to use.

Green: Colin, have you also tried the reaction using unsubstituted butadiene?

Suckling: We've not done that. All the substrates have one X substituent—methyl, methoxy or acetoxy.

Lerner: Does the sequence of addition of the reactants make any difference?

Tedford: I have always added the proposed inhibitor first to make sure that the antibody was saturated, before adding reactants.

Schwabacher: Are these reactions always completely homogeneous?

Tedford: Yes. The reactions are carried out in aqueous solution containing less than 2% acetonitrile. We use quite dilute solutions, 4.7×10^{-4} M.

Schultz: How do you assay the reaction?

Suckling: By HPLC.

Schultz: Did you observe any hydrolysis of the esters?

Tedford: There is a small amount of hydrolysis of the acetoxybutadiene.

Blackburn: Some remarkable rate accelerations have been achieved in Diels–Alder reactions by using very high salt concentrations. These have been compared to the effect of very high pressure. Has anyone looked at the effect of salt concentration on the uncatalysed Diels–Alder reaction? Could this mimic what the antibodies are doing?

Hilvert: The uncatalysed version of the Diels–Alder reaction we studied occurs much faster in water than in organic solvents. Although we include 100 mM sodium chloride in the aqueous reaction, the salt concentration is unlikely to be significant.

Schultz: The salt effects reported are seen with concentrations of lithium chloride above one molar.

Professor Suckling, have you looked at the energy minimized structure of the cyclohexene product?

Suckling: We have.

Benkovic: There are two pK_as in the reaction profile for the catalysis of the Diels–Alder reaction by antibody H11 (Fig. 2).

Suckling: Yes. I would like to extend the pH studies to a more alkaline pH, to verify this.

Janda: How long do you run the reactions for? Does the antibody get degraded at all?

Tedford: The reactions are run for about an hour to an hour and a half. We have followed it over pH 4–9.5 and seen no evidence of antibody degradation.

Schwabacher: The acetoxy butadiene may be decomposing fairly rapidly at that pH. That could be why the rate appears to fall.

Plückthun: For the nitrophenyl acetate hydrolysis, the pH rate profile (Fig. 4) was determined under saturating substrate conditions, so it really reflects the change in k_{cat}.

Lerner: With catalytic antibodies, because you have induced the pocket, you know something about it. With enzymes, unless you solve the crystal structure, you know virtually nothing about the structure of the active site. In enzymology you often have kinetic evidence before you have any feel for the structure of the pocket. In catalytic antibodies the reverse is true. That means there are many interesting things to do as long as there are substrate opportunities.

Suckling: We have a number of antibodies with quite interesting structures. One binds nicely to a toxin that contains a special type of alkyne; there might be some interesting chemistry there. We have antibodies that bind to different α-toxins, which present several opportunities for selective reactions. There is a lot that can be looked at.

Paul: The moral of the story for me is that there are many unknown factors involved here. The immune system is not fully understood and we appear to have only scratched the surface in understanding catalysis by antibodies.

Expanded transition state analogues

G. Michael Blackburn*, Gillian Kingsbury†, Sriyani Jayaweera* and Dennis R. Burton†

*Departments of *Chemistry and †Molecular Biology, Krebs Institute, Sheffield University, Sheffield S3 7HF, UK*

Abstract. Stable analogues of transition states are used as haptens to elicit antibodies that will catalyse the reaction under investigation. The present failure of such antibodies to equal the catalytic efficiency of enzymes has prompted a more detailed analysis of the structure of transition states. Calculations suggest that for many types of reaction, including the Diels–Alder reaction and addition–elimination reactions, interatomic distances in the transition state are longer than those in analogues hitherto used to elicit catalytic antibodies. Three possible solutions are proposed for the design and synthesis of stable analogues that match the longer interatomic distances of transition states: atom substitution, atom insertion and double atom insertion. Some applications of these ideas to specific reactions are described. The concept of expanded transition states is also valuable for associative processes as a way of avoiding product inhibition of the catalytic antibody; this is being explored for a phenylalanine transcarbamylase antibody.

1991 Catalytic antibodies. Wiley, Chichester (Ciba Foundation Symposium 159) p 211–226

Transition state theory has achieved widespread acceptance as a tool for the understanding of many reaction rate data and it has provided useful insights into a variety of chemical and physical processes (Laidler & King 1983). Two of the fundamental assumptions in transition state theory are that molecules must traverse saddle points (transition states) of the potential energy surface and that rates of reactions are proportional to the concentrations of molecules in the transition state, which, in turn, must be in quasi-equilibrium with reactants. It follows that the nature of the transition state is the focus of the entire theory.

In a similar fashion, the theory behind the performance of catalytic antibodies is espoused in the original statements of Pauling (1948) who said: 'Enzymes are molecules that are complementary to the structure of the activated complexes of the reactions that they catalyse'. That concept was tied into the language of transition states by Jencks (1969), who argued that antibodies raised against a synthetic mimic of a transition state might provide proteins with enzyme-like catalytic properties.

In the exciting discoveries of the last four years, antibodies have been identified that have the ability to catalyse a wide range of reactions. Considerable ingenuity has been displayed in the design of stable analogues of transition states and these have been employed as haptens to elicit expression of the desired antibodies (Schultz et al 1990). However, the performance of these protein catalysts still leaves much scope for improvement if they are to match the catalytic power of most enzymes. The purpose of this paper is to enquire whether our approaches to the design of transition state analogues might be assisted by consideration of existing knowledge regarding the structure of transition states for a range of chemical processes, including nucleophilic substitution reactions at unsaturated and saturated centres, cycloaddition processes and intramolecular rearrangements. That in itself begs the question of whether antibodies need even be raised to transition state analogues. One must ask whether the success of Schultz's induction of a carboxylate in the active site of an antibody to effect a β-elimination reaction (Shokat et al 1989) is an indication of a more profitable approach to catalytic monoclonal proteins. In the light of the Leffler–Hammond postulate—which states that for reactions with transition states intermediate between reactants and products, factors which stabilize these features in the products will also do so in the transition state relative to the starting materials (Jencks 1969, Leffler & Grunewald 1963)—bisubstrate analogues might be a better choice as haptens, especially for reactions having 'early' transition states.

The deciphering of transition state structures in chemical reactions has been, and remains, a major goal in mechanistic chemistry. Only recently has a direct spectroscopic method been devised for the observation of a transition state (Collings et al 1987). In general, their structures are inferred from quantitative analysis of rate processes, particularly linear free energy relationships and kinetic isotope effects, while our understanding of transition state geometry and charge distribution has developed particularly well from the recent application of a range of computational methods to model systems, inevitably simplified to limit the number of atoms and/or electrons involved in the calculations.

Because a major challenge for catalytic antibodies is the facilitation of reactions that lie outside the domain of enzyme catalysis in type, selectivity, or stereospecificity, it is essential that we broaden our understanding of the 'rules' that govern the performance of catalytic antibodies so that their development does not remain locked onto a few select examples whose characteristics are propitious for the exploitation of Fab characteristics. It is therefore profitable to ask 'What are the spatial details of transition state geometry?' in order that we can improve the design of stable analogues.

Diels–Alder cycloaddition

While there is speculation that the Diels–Alder cycloaddition may occur spontaneously in a few biosynthetic pathways for secondary metabolites, notably for intramolecular examples, there is no obvious enzyme candidate for a natural 'Diels–Alderase'. Yet the current deployment of this reaction for synthetic utility is widespread. What are the characteristics of the transition state for such processes? The simplest example, the reaction of butadiene with ethylene, is a synchronous concerted reaction whose C_s symmetrical transition state can be reached by a variety of calculations (Houck et al 1989). Suggestions that asymmetric substitution might cause the transition state structures to become asynchronous and involve diradical intermediates appear to be unfounded, at least for the addition of cyanoethylenes to butadiene and to cyclopentadiene. Houck's *ab initio* calculations at the 3-21G level (Table 1) give the optimized geometries of the transition structures for cyclopentadiene addition to acrylonitrile (Fig. 1). There are two significant features: first, long interatomic distances between the newly joining carbon atoms (in the range 2.13 to 2.29 Å) and, secondly, little distortion from trigonal geometry at these four carbon centres.

TABLE 1 3-21G calculated and experimental activation energies for the addition of cyclopentadiene to acrylonitrile

	Bond length in transition state (Å)			
Stereochemistry	*C1····C2*	*C3····C4*	$\Delta E-10$	E_a *(kcal/mol)*
endo	2.29	2.13	16.0	13.2
exo	2.29	2.13	15.7	13.4

This poses seemingly insuperable difficulties for the accurate design of a stable transition state analogue for a bimolecular Diels–Alder process, since the Periodic Table does not contain atomic species with the right characteristics for the purpose! *Faut de mieux*, one might (a) employ a product-like hapten in conjunction with special devices to overcome the seemingly inevitable inhibition of the catalytic antibody by the Diels–Alder product; (b) fix the juxtaposition of non-bonded partners in a bisubstrate analogue, using an intramolecular bridge so that the whole occupies a molecular volume equivalent to that to be adopted by appropriate substituents in the reactants; or (c) augment an analogue of a 'late' transition state by additional atoms bonded to its *outside*, again with the net effect of filling a volume equivalent to that of an 'early' transition state. While the first approach has been nicely exemplified by Hilvert's cycloaddition process (Fig. 2), which avoids the inhibition trap through the spontaneous extrusion of SO_2 from the product (Hilvert et al 1989), it does not appear to provide a general solution to the problem, which may require a detailed exploration of bisubstrate analogues as haptens for Diels-Alderase antibodies.

FIG. 1. Addition of cyclopentadiene to acrylonitrile.

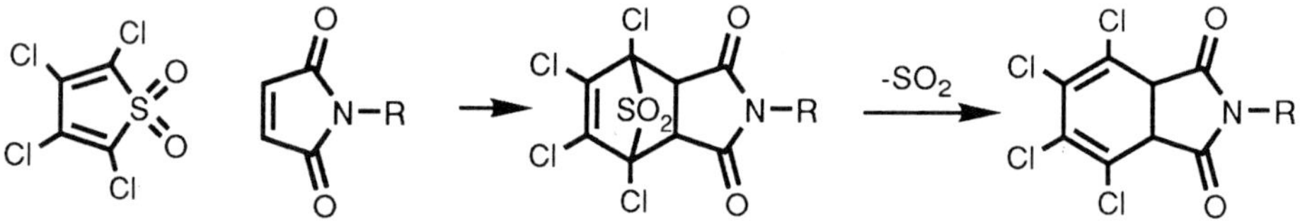

FIG. 2. Cycloaddition followed by expulsion of SO_2 allows a Diels–Alder reaction to proceed without product inhibition of the catalytic antibody.

Addition–elimination reactions

These reactions, especially for unactivated alkenes, provide a serious target for catalytic antibody development. E2 processes occur in one step with favoured *anti*-geometry, and have been studied by a variety of computational methods. A recent study on the E2 elimination of ethyl chloride includes structural analysis of early and late transition states, both of which are *anti* and E2H like (Minato & Yamabe 1988). For the early transition state, where the base is F^-, the C····Cl distance is 2.29 Å, while for the late transition state, with the weaker base Cl^-, the C····Cl distance is 2.74 Å (Fig. 3A). It is suggested that solvation effects would be largely electrostatic and not make major changes to these geometries. Clearly, the mimicry of such long interatomic distances in transition state analogues calls for new strategies if antibodies are not to be restricted to the catalysis of only E1-like elimination processes.

Aldol condensation

Of the non-concerted addition reactions, the aldol condensation has great practical utility and offers tremendous scope for stereochemical control. Ken Houck has computed transition state structures for reactions of formaldehyde with acetaldehyde enolate and its lithium and borate derivatives (Li et al 1988). The shortest C1····C2 interatomic distance is for the naked enolate anion and is close to 2.0 Å, irrespective of the conformation (*gauche* or *anti*), whereas it approaches 2.4 Å for the lithium enolate with an earlier transition state (Fig. 3B).

Hydrolysis of acetals

Using a computational approach to the uncatalysed hydrolysis of acetals based on extrapolation from ground-state structures using valence force constants plus energies of activation, Bürgi has deduced that the transition state for cleavage of an α-hexopyranoside has a C1····O interatomic distance of 1.6 to 3 Å (Bürgi & Dubler-Steudle 1988). Clearly, an analogue to elicit a glycosidase abzyme may need to mimic a very long bond indeed.

Substitution reactions

These present a major challenge to scientists working with catalytic antibodies: first to supplement the very small number of enzymes that catalyse single-step S_N2 processes and, secondly, to overcome the problem of sequential transition states for two-step acyl (and phosphoryl) group transfers.

Shi & Boyd (1989) have completed a 6-31G calculation of the asymmetric transition state for an S_N2 displacement reaction involving hydroxide and methyl chloride. The transition state is early, with a 2.66 Å separation of C1····O, charge distribution related to the C—O and C—Cl bond lengths, and clear tetrahedral geometry at carbon.

Carbonyl addition reactions

When, lastly, one turns to heteroatom additions to the solvated carbonyl group, the early geometrical analysis of the direction of attack of Bürgi & Dunitz has been developd by Karplus using the hypernetted chain extended reference interaction site model integral equation theory for the addition of hydroxide to formaldehyde with particular attention to the role of water (Yu & Karplus 1990). The C1····O2 distance in the transition state shifts from 2.39 Å in the gas phase to 2.0 Å. Activation energies are in approximate agreement with those determined experimentally. In the transition state there is still more negative charge on the attacking hydroxyl oxygen than on the carbonyl oxygen. Related results have been found for the water-assisted addition of ammonia to formaldehyde (Williams 1987), which involves a cyclic transition state for a two-water-molecule system that is clearly tetrahedral at C1 *and* nitrogen, with an N····C1 bond distance of 1.59 Å. Such results support the focus of attention on tetrahedral geometry for transition state analogues of acyl transfer processes but still leave scope for improved design with regard to interatomic distances and to asymmetric charge distribution.

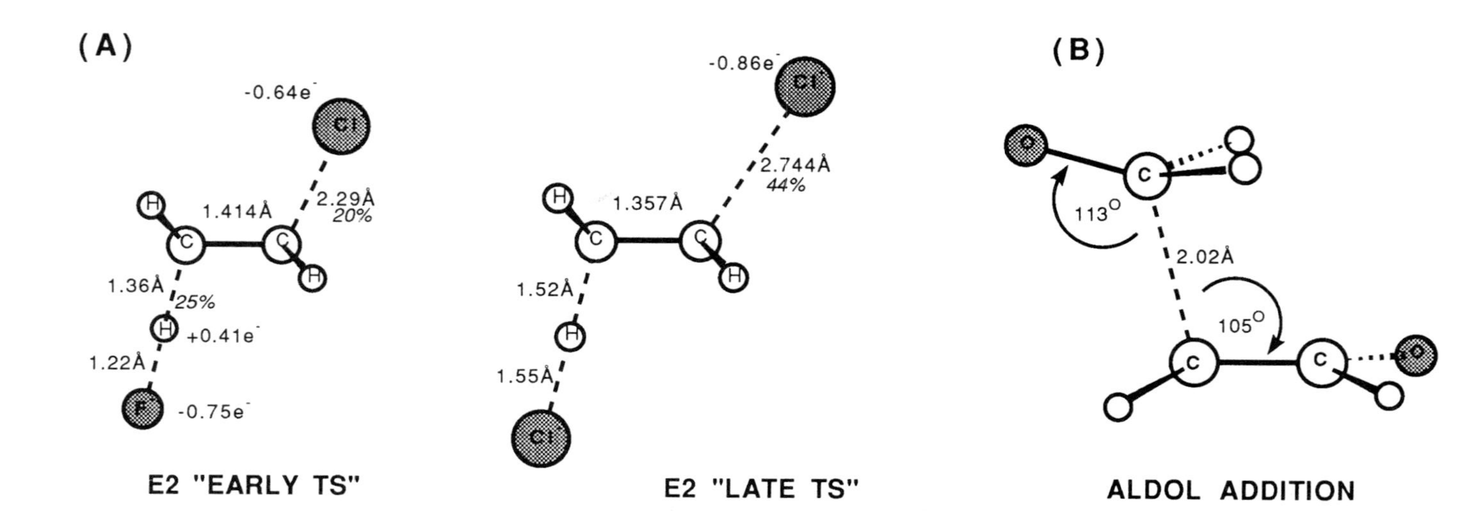

FIG. 3. The transition state geometries of (A) an addition–elimination reaction and (B) an aldol condensation.

Improved transition state analogues

These and other considerations have prompted us to direct attention to two particular features of the design of transition state analogues, namely:

(i) the simulation of longer interatomic distances in 'stretched transition state analogues',

(ii) the avoidance of product inhibition for reactions with late transition states.

It seems to us that there might be three solutions to the problem of lengthening bonds in transition state analogues to match the long interatomic distances (often in excess of 2 Å) that have been identified for early transition states through calculations of the varieties described above. These are:

(a) Atom substitution, exemplified by the design of an analogue containing a P—S—C fragment to mimic a C—O—C component, as in a tetrahedral transition state for acyl transfer;

(b) Atom insertion, illustrated by the simulation of a long C—O—C in a transition state by a P—X—O—C surrogate, where X might be CH_2 or NH;

(c) Double atom insertion, where a long C—O—C fragment might be adequately replaced by a P—CX=CX′—C with X = H or F to provide the lone-pair characteristics of the prototype oxygen atom.

As an example of atom insertion we are seeking to develop Schultz's phosphonate hapten (**1a**) (Fig. 4) through replacement of oxygen by sulphur. The resulting analogues either have changed geometries, as illustrated for the thiolphosphonate (**1b**) or desymmetrized charge distribution (**1c**). In particular, the P—C1 bond length increases from 2.5 Å in **1a** to 3.0 Å in **1b**. The use of all these species as transition state analogues in a competitive fashion should permit the comparison of their relative effectiveness through the reactivity of the best catalytic antibodies for the hydrolysis of ethyl 4-nitrophenyl carbonate (**1d**) (Jacobs et al 1987) by application of the λ Fab technology (Huse et al 1989).

The atom insertion approach can be applied to the aminolysis of an ester, for example the reaction of piperazine with 4-nitrophenyl acetate (PNPA). The anionic tetrahedral intermediate for this process (**2a**) (Fig. 5) should break down more readily than it is formed, identifying the transition state for the process as piperazine added to PNPA. Thus, the design of an analogue for an early transition state should emphasize the long (O)C····N bond distance, as illustrated for **2b**, where the insertion of either a methylene group or an oxygen atom between the tetrahedral phosphorus and the piperazine N^{α} will provide a stretched structure with an increase of 0.9–1.2 Å over the phosphinamide (Fig. 5).

Similar atom insertion design can be applied to transition state analogues for E2 elimination processes or additions of nucleophiles to conjugated or unconjugated C=C double bonds as, for example, in the ketone **2c**.

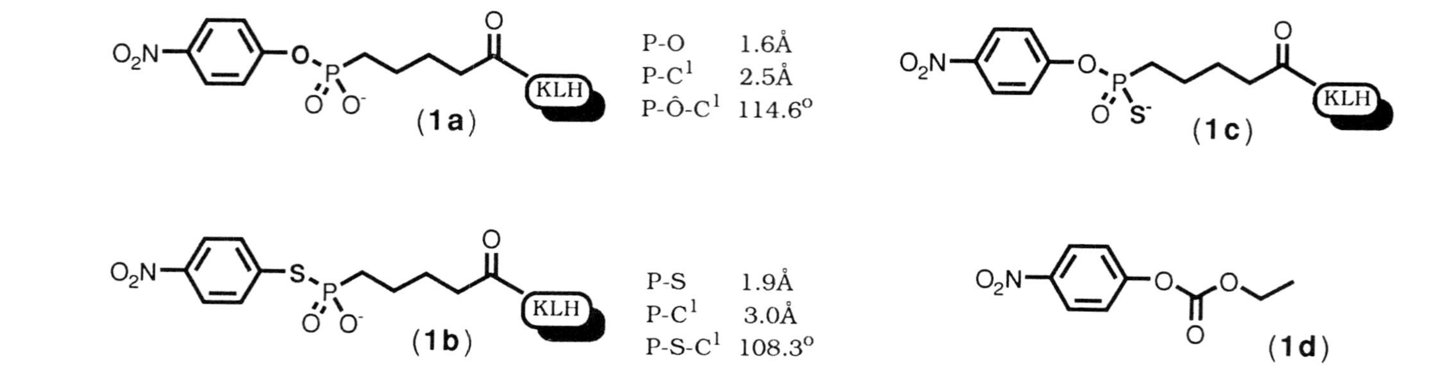

FIG. 4. Structures of transition state analogues **(1a–c)** used to raise antibodies to catalyse the hydrolysis of ethyl 4-nitrophenyl carbonate **(1d)**. KLH, keyhole limpet haemocyanin.

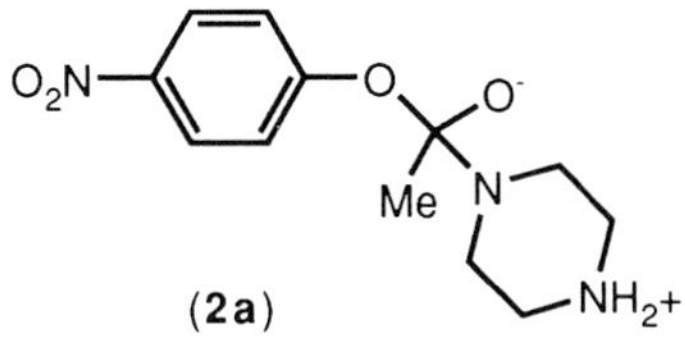

O_2N O O^- Me N NH_2+ (**2a**)

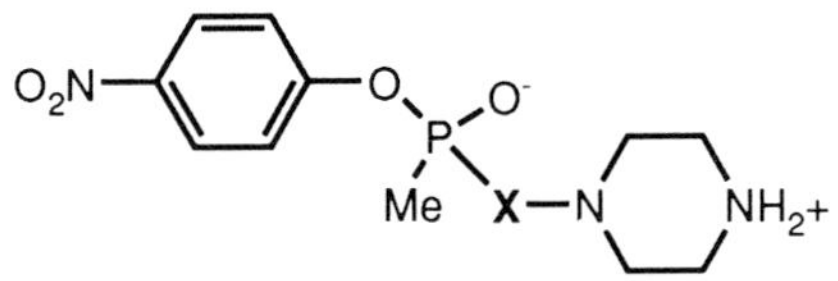

O_2N O P O^- Me X—N NH_2+ (**2b**)

O O NH_2 H H_2N^+ (**2c**)

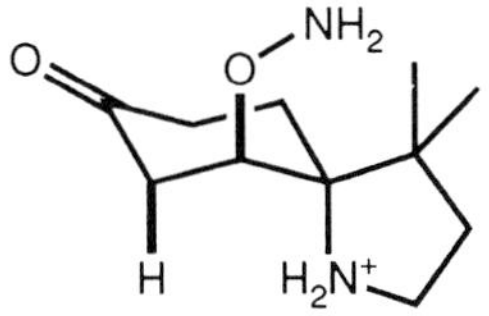

Bond Geometries for (**2b**)

	X=(-)	**X**=CH_2	**X**=O
P-N	1.5Å	2.7Å	2.4Å
O^--P-N	114°	140°	133°

FIG. 5. Structures of the transition state (**2a**) and an analogue (**2b**) for the reaction of piperazine with 4-nitrophenyl acetate. (**2c**) Transition state analogue for addition of water to a cyclohexenone.

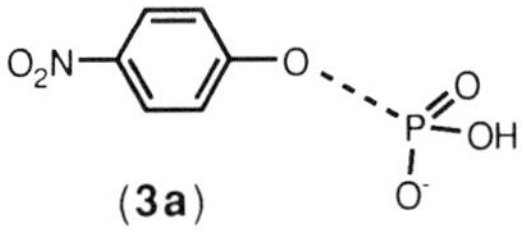

O_2N O P O OH O^- (**3a**)

O_2N F α PO_3H^- F

O_2N F α F PO_3H^-

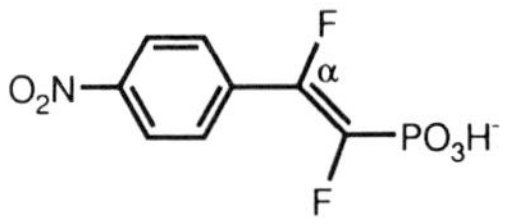

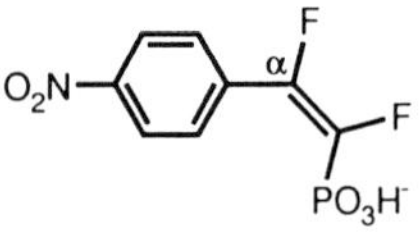

Difluorostyrylphosphonate		*trans*	*cis*
(**3c**)	ρ P-C^{α}	2.2Å	2.2Å
	P-C^{α}-C^1	148.5°	104°

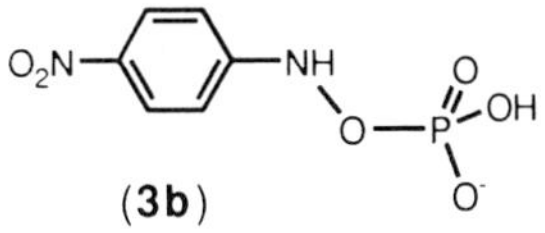

O_2N NH O—P O OH O^- (**3b**)

O_2N F F P O OH O^- 2.2Å

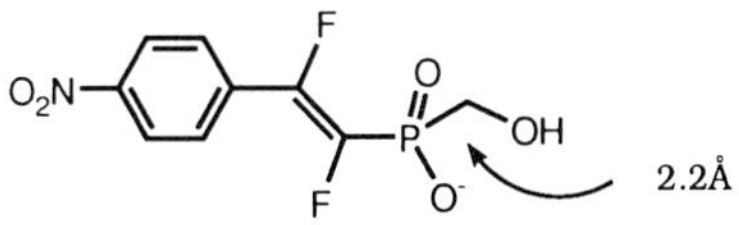

"Exploded TS" Analogue for Phosphate Hydrolysis (**3d**)

FIG. 6. Stretched transition state analogues (**3b–d**) for the hydrolysis of 4-nitrophenyl phosphate monoanion (**3a**).

The task of analogue design for phosphate ester hydrolysis for an associative process (Herschlag & Jencks 1989) seems to call for a stable pentacoordinate species where there is a limited range of elements, e.g. vanadium, of doubtful stability. By contrast, hydrolysis of a phosphate monoester through a dissociative process can be attempted through application of the concept of stretched transition state analogue design. The anticipated long P—O bond distance for the hydrolysis of 4-nitrophenyl phosphate monoanion (**3a**) (Fig. 6) could be

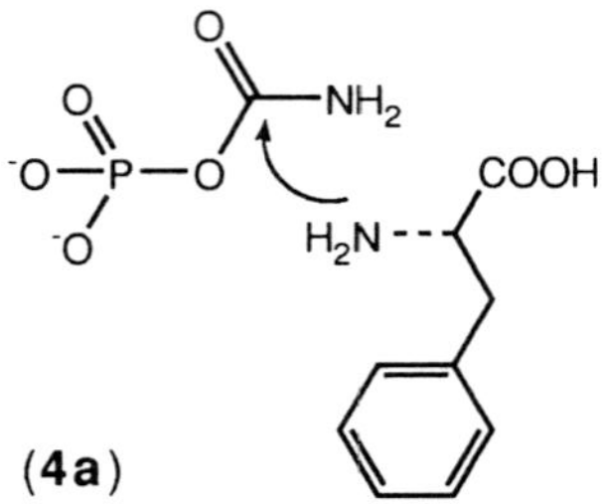

(**4a**)

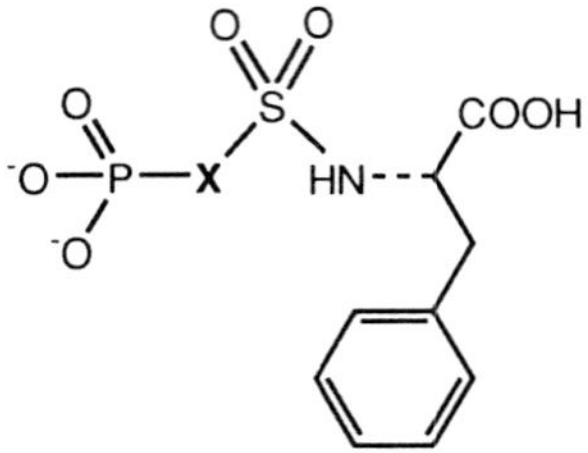

(**4b**)

X = CH_2, CH=CH, *etc*

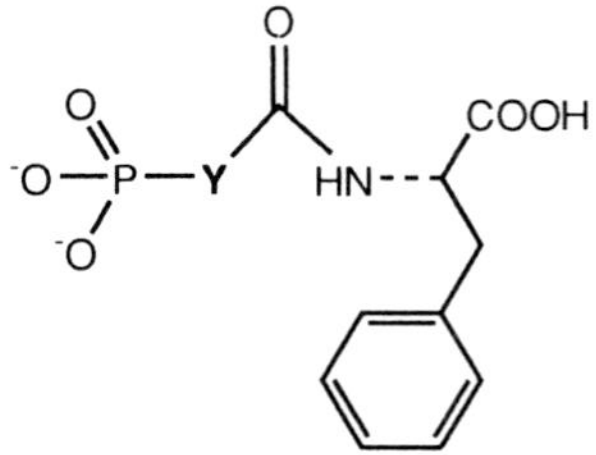

(**4c**)

Y = CH_2, CH=CH, CF_2, OCH_2, *etc*

FIG. 7. Transition state analogues (**4b–c**) for the reaction between phenylalanine and carbamyl phosphate (**4a**).

simulated either by atom insertion, employing an *N*-phenyl-*O*-phosphoryl-hydroxylamine (**3b**), or by double atom insertion through the use of β-styrylphosphonic acids. The lone-pair electrons on the leaving oxygen might be identified through the substitution of one or both of the styryl α- and β-hydrogens by fluorines (**3c**). Mimicry of the incoming nucleophilic water in the exploded transition state can be achieved through the use of a (hydroxymethyl)(styryl)phosphinic acid (**3d**).

Finally, the concept of expanded transition states can be applied to the design of analogues for synthetically useful, associative processes. For the simple reaction, $\mathbf{A}+\mathbf{B}\rightarrow\mathbf{AB}$, product inhibition of a catalytic antibody seems almost guaranteed because of the inevitable similarity between the features of $[\mathbf{A.B}]^{\ddagger}$ and **AB**. However, the metathetical process, $\mathbf{A}+\mathbf{B}\rightarrow\mathbf{C}+\mathbf{D}$ can avoid that trap, provided that the (unwanted) dissociating fragment **D** does not bind strongly to the catalytic antibody. We are exploring this system for a phenylalanine transcarbamylase antibody based on the proven success of phosphonoacetyl-L-aspartate (PALA) as a powerful inhibitor for aspartyl transcarbamylase (Stark & Collins 1971). The transition state is expected to be formed by attack of the amino group of phenylalanine on carbamyl phosphate (**4a**) (Fig. 7). This is being mimicked by sulphonamides with variable spacers between the phosphoryl and sulphonyl centres (**4b**) and by acetamides, also with variable geometry (**4c**), which more closely resemble PALA. Expanded transition state analogues could also be formed by atom insertion between the phenylalanine and the sulphonyl tetrahedral centre. Linkage of **4b** ($X=CH_2$) and of **4c** ($Y=CF_2$, CH_2) to the carrier proteins, bovine serum albumin or keyhole limpet haemocyanin, has been readily achieved through a *p*-amino-substituent on the phenylalanine ring.

We believe that the pursuit of these two objectives will provide significant understanding of the rules that govern the performance of catalytic antibodies.

Acknowledgements

We thank the SERC (Protein Engineering and Biotransformation Initiatives) and the University of Sheffield for support of this work.

References

Bürgi H-B, Dubler-Steudle KC 1988 Empirical potential energy surfaces relating structure and activation energy. 2. Determination of TS structure for the spontaneous hydrolysis of axial tetrahydropyranyl acetals. J Am Chem Soc 110:7291-7299

Collings BA, Polanyi JC, Smith MA, Stolow A, Tarr AW 1987 Observation of the transition state dideuterium monohydrogen dipositive ion (HD_2^{2+}) in collisions, atomic hydrogen + molecular deuterium. Phys Rev Lett 59:2551–2554

Herschlag D, Jencks WP 1989 Phosphoryl transfer to anionic oxygen nucleophiles. Nature of the transition state and electrostatic repulsion. J Am Chem Soc 111:7587–7592

Hilvert D, Hill KW, Nared KD, Auditor M-TM 1989 Antibody catalysis of a Diels–Alder reaction. J Am Chem Soc 111:9260–9261

Houk KN, Loncharich RL, Blake JF, Jorgensen WL 1989 Substituent effects and transition structure for Diels–Alder reactions of butadiene and cyclopentadiene with cyanoalkenes. J Am Chem Soc 111:9172–9176
Huse W, Sastry L, Kang AS et al 1989 Generation of a large combinatorial library of the immunoglobulin repertoire in phage lambda. Science (Wash DC) 246: 1275–1281
Jacobs J, Schultz PG, Sugasawara R, Powell M 1987 Catalytic antibodies. J Am Chem Soc 109:2174–2176
Jencks WP 1969 Catalysis in chemistry and enzymology. McGraw-Hill, New York, p 193 & 288
Laidler KJ, King MC 1983 The development of transition state theory. J Phys Chem 87:2657–2661
Leffler J, Grunewald E 1963 Rates and equilibria of organic reactions. Wiley, New York, p 156
Li Y, Paddon-Row MN, Houk KN 1988 Transition structures of aldol reactions. J Am Chem Soc 110:3684–3686
Minato T, Yamabe S 1988 Comparative study of E2 and S_N2 reactions between ethyl halide and halide ion. J Am Chem Soc 110:4586–4593
Pauling L 1948 Nature of forces between large molecules of biological interest. Nature (Lond) 161:707–709
Schultz PG, Lerner RA, Benkovic SJ 1990 Catalytic antibodies. Chem Eng News 68:26–40
Shi Z, Boyd RJ 1989 Transition-state electronic structures in S_N2 reactions. J Am Chem Soc 111:1575–1579
Shokat K, Leumann CJ, Sugasawara R, Schultz PG 1989 A new strategy for the generation of catalytic antibodies. Nature (Lond) 338:269–271
Stark GR, Collins KD 1971 Aspartate transcarbamylase: interaction with the transition state analogue *N*-(phosphonoacetyl)-L-aspartate. J Biol Chem 246:6599–6605
Williams IH 1987 Theoretical modelling of specific solvation effects upon carbonyl addition. J Am Chem Soc 109:6299–6304
Yu H-A, Karplus M 1990 An integral equation theory study of the solvent-induced reaction barrier in the nucleophilic addition of hydroxide to formaldehyde. J Am Chem Soc 112:5706–5716

DISCUSSION

Lerner: If one wants to mimic the attack of water on the carbonyl carbon, would it help to put an NH_2 mimic on the phosphorus—assuming it would be stable?

Blackburn: Yes. That is the equivalent to the structure for the E2 process in reverse (Fig. 5, **2c**). The unsubstituted hydroxyl then advantageously has about the right p*K*, although one doesn't know the state of protonation of that group when it's inside the protein site.

Schultz: One thing we have ignored at this meeting is hapten synthesis. In some cases hapten structure is dictated to some degree by the ease of synthesis. In our Diels–Alder reaction the bond lengths were rather short, but to generate a structure with the correct bond lengths is not easy.

The calculations you described do not include configuration interaction. For transition states that are open shell configurations, one must use extensive configuration interaction.

Blackburn: I accept that, but I don't think we are concerned with absolute energies in this process. Also, the calculated geometries are likely to be more accurate than the absolute energies.

Schultz: The relative energies of various transition states are very hard to calculate accurately; for example, the bent Diels–Alder transition state relative to a symmetrical one.

Blackburn: But in general the development of tetrahedral character at the joining bonds is not far advanced.

Schultz: I agree with that. There are better transition state analogues, but they are synthetically more challenging.

Page: You are right to question the models that are used for transition states but I am not sure that using theoretical calculations is a good way to justify that. The calculations you quoted minimize enthalpy and they don't minimize the entropy. Solvent effects are very important in determining the configuration of the transition state. The few calculations that have included water molecules (and again it's usually only one or two, so you end up with cyclic structures) tend to give different bond lengths than do calculations for those structures in the absence of solvent.

Jencks: Generally, for a transition state of a certain dimension made from two spherical molecules, does one want to make a site that will fit that transition state exactly and provide good stabilization? Or does one want to make a site that is somewhat smaller than the transition state, so that for the two separate molecules to get into it they have to be compressed and a force exerted on them? This would provide strain in the enzyme–substrate complex that would be released in the transition state. It may also provide larger force constants for compression and for loss of energy of the bound substrate.

Schwabacher: It might depend on whether you are trying to get the best k_{cat} or the best k_{cat}/K_m. If you want to bind the transition state most strongly (k_{cat}/K_m), presumably you want the binding site to be exactly the right size for it. But if you want to distinguish optimally between transition state and substrate (k_{cat}), you might want the binding site to be smaller.

Hansen: When the pocket is smaller, won't it look more like product? If you are thinking about two pieces coming together, you want the pocket that exactly fits the transition state. The two pieces in the ground state still have to be squeezed to fit into the transition state pocket as the bond forms and then the product would be smaller than the pocket.

Jencks: The site is not completely rigid and it does have to exert a force to bring the bound substrate into the structure of the transition state.

Lerner: Where is the shrinking taking place? Is it taking place in the solution or in the antibody pocket? If the lifetime of the transition state is that of a bond vibration and the binding is diffusion controlled, that would exclude any sort of trapping or solution effect.

Jencks: Presumably, the molecules bind before they reach the transition state.

Page: Basically, binding energy is needed to compensate for the loss of translational and rotational entropy that occurs when the two separate entities bind. The loss of entropy required for covalent bond formation could occur on initial binding. Then as long as there are sufficient degrees of freedom left in the reactants so that they are not unnecessarily constricted, no more entropy loss would occur in the transition state. The loss of entropy is not necessarily related to a particular geometry. You don't need an exact complementary geometrical fit in one particular state—neither the ground state nor the transition state—to compensate the entropy loss.

Jencks: Certainly, the loss of rotational and translational entropy is very important. But it is difficult to achieve that loss of entropy. The question is to what extent an enzyme can freeze motions of bound molecules. For maximum loss of entropy it is necessary to hold these molecules very strongly. The most important reason for a really tight transition state, and what would normally be called steric strain, is likely to be loss of entropy and a thorough freezing of the molecules. The entropy of a crystalline solid is enormous; it is not very different from that of the molecule in solution.

Lerner: One concern is to what extent a 20% change in the bond length will be read in a realistic way by the antibody. Given the subtlety of the recognitory events, there is hope that these small changes in geometry will be recognized.

Hansen: Transition state analogues are used as enzyme inhibitors. Can one get clues about what structures are really good transition state mimics from their relative potencies as enzyme inhibitors?

Blackburn: Yes. Studies of phosphonate analogues of substrates for various enzymes show that some of these break all the rules. There are attenuated phosphonates where the oxygen between the carbon skeleton and the phosphorus atom is simply deleted (Freeman et al 1989, Widlanski et al 1989). These analogues bind much more tightly to EPSP synthase than do the isosteric phosphonates, which require a tedious synthesis to introduce the CH_2 group. These non-isosteric molecules don't do anything other than act as inhibitors. Jeremy Knowles' attenuated inhibitor shows no deuterium exchange in this cyclohexanol methylphosphonate system. But when you put the extra carbon in place of the phosphate oxygen, the inhibitor fits the enzyme site less tightly but more appropriately for the reaction coordinates. So the enzyme can catalyse the redox process, exchange deuterium and go back again. Those examples say that enzymes are sensitive to missing atoms or to quite small bond distances.

Lerner: Given the probability of failure, who is going to do the 25-step syntheses necessary to produce some of these transition state analogues? Will it be industry trying to do an important commercial reaction? Or will it be graduate students, and how do they get their degrees if the synthesis doesn't work?

Blackburn: I regard my colleagues in synthetic chemistry with a mixture of respect for their ability and despair for their choice of targets! Organic chemistry in my view is ripe for an incursion by people who set new challenges. One great target for that process is the definition of weak intramolecular forces that are centred on this project.

Lerner: Rappoport said the most interesting thing about catalytic antibody research is that it offers new targets for synthetic chemistry.

Hilvert: But we may already know something about whether or not the antibody can read a few tenths of an Ångström difference in bond lengths. Presumably, the relative affinities of individual antibodies to aromatic compounds with different halogen substituents is known.

Benkovic: There are tables of such data for binding of antibodies to fluoro, bromo and chloro derivatives.

Plückthun: Asking whether there is always an antibody in the repertoire that will recognize one particular compound better than it recognizes another one presumes that you can put an antibody atom at any given place, for example, as a side chain on a loop. The variation in the repertoire is enormous but not infinite. There are certain points in space where there is probably no way that there can be an atom supplied by the antibody.

Benkovic: Michael Blackburn's compounds may be less susceptible to product inhibition, because he is creating a larger pocket. I agree with Bill Jencks that there is the problem of getting substrates in very close by sacrificing translational and rotational entropy. On the other hand, antibodies in general are not going to dissociate products efficiently unless we exaggerate the steric or electronic differences in product molecules. A series of analogues that not only considers the charge distribution in the transition state, but also exaggerates the electronic differences that occur once the product is formed, may be more successful. We are starting to look at the kinetics of product binding. Will we obtain the conformational changes in the hypervariable loops required to release the reaction products—especially in bimolecular reactions?

Jencks: Substitution reactions are not a problem.

Benkovic: Yes, but an antibody-catalysed reaction that brings substrates together in a condensation reaction may present difficulties with product desorption.

Janda: A lot of people are concerned with bond lengths and angles, but we find in a number of cases that if one immunizes with something specific or uses an antibody raised against a specific small molecule, a number of compounds are being accepted in the binding pocket of the antibody that we wouldn't expect to bind there.

I started working on catalytic antibodies in 1985 and I was very concerned with the overall design of the haptenic molecules (i.e. using close mimics for bond angles and charges). But we have backed off from that and we are looking at something that is not quite as difficult. We are using molecules that may not match everything that is needed in the transition state and hoping that the immune repertoire will provide what we want.

Lerner: You can't impose a style on science. Everyone has different purposes. People who want to make commercially important compounds will get every bond length and every charge correct, because the pay off is reasonable for them.

Schwabacher: But the question is whether making the transition state analogue resemble the true transition state that much more accurately will really produce a better antibody.

Lerner: It is very important to find out.

Schultz: Even with enzymes, there are only a few cases, such as thermolysin, where transition state mimicry has been determined unambiguously (Bartlett & Marlowe 1983).

References

Bartlett PA, Marlowe CK 1983 Phosphonamidates as transition-state analogue inhibitors of thermolysin. Biochemistry 22:4618–4624

Freeman S, Seidel HM, Schwalbe CH, Knowles JR 1989 Phosphonate biosynthesis: the stereochemical course of phosphoenolpyruvate phosphomutase. J Am Chem Soc 111:9233–9234

Widlanski T, Bender SL, Knowles JR 1989 Dehydroquinate synthase: a sheep in wolf's clothing? J Am Chem Soc 111:2299–2300

Tritylase antibodies

Brent L. Iverson*, Sheila A. Iverson*, Katherine E. Cameron, Guiti K. Jahangiri, Dahna S. Pasternak and Richard A. Lerner

Department of Molecular Biology and Chemistry, The Research Institute of Scripps Clinic, 10666 North Torrey Pines Road, La Jolla, CA 92037, USA

Abstract. We have used a tris(4-methoxyphenyl)-phosphonium compound as a hapten to elicit catalytic antibodies that selectively remove trityl protecting groups at neutral pH. One antibody, 37C4, was characterized kinetically with a number of trityl substrates. The rate enhancement was consistently near 200; the K_m was approximately 30 μM for the methoxytrityl substrates. Compounds with no methoxy substituents on the trityl group were not hydrolysed by the antibody. No decrease in the rate of reaction was detected through 21 turnovers, which suggests that the presumptive trityl cation formed during the cleavage reaction does not alkylate the antibody binding pocket. The rates of the background and antibody-catalysed reactions both increase logarithmically with decreasing pH, implying that general acid catalysis is not involved: further studies will test this assumption. The favoured mechanism for the catalytic activity of antibody 37C4 is charge complementarity in the binding site stabilizing a positively charged intermediate(s) in the cleavage reaction. The coding sequence for 37C4 is being cloned into a phage λ vector in preparation for site-directed mutagenesis to improve the catalytic efficiency of the antibody.

1991 Catalytic antibodies. Wiley, Chichester (Ciba Foundation Symposium 159) p 227–235

Catalytic antibodies have now been produced that are capable of effecting a remarkable array of chemical transformations (Lerner & Benkovic 1990, Schultz et al 1990). A unique advantage of antibody catalysts is that profound substrate specificity can be programmed into the catalyst through appropriate selection of hapten structure. An ideal catalytic antibody would be capable of bringing two complex molecules together to carry out a desired specific transformation, regardless of other modes of reaction which may be possible in the absence of antibody catalysis. Such a system would obviate the need for chemical protecting groups. However, if this goal is not achieved and protecting groups are still

*Present address: Department of Chemistry, The University of Texas at Austin, Austin, TX 78712, USA.

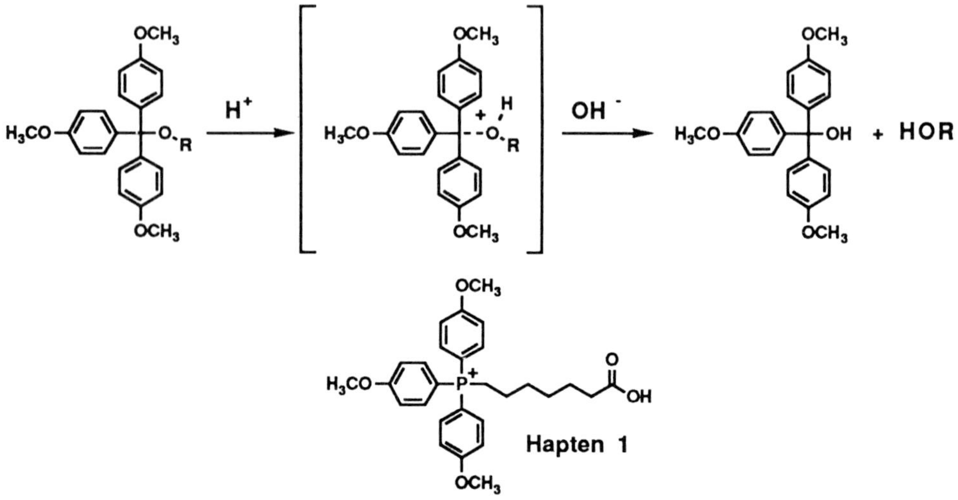

FIG. 1. Structure of phosphonium hapten **1** and outline of the reaction catalysed by monoclonal antibody 37C4.

required, then the programmable substrate selectivity of antibodies could be useful in the manipulation of those protecting groups. Structurally distinct yet chemically similar protecting groups can be envisioned, with corresponding catalytic antibodies being used to achieve a desired reaction sequence. Antibody catalysis is well suited to this strategy because the mild conditions employed would be compatible with even sensitive chemical functionalities.

We have produced an antibody catalyst which selectively removes trityl protecting groups at neutral pH (Iverson et al 1990). A tris(4-methoxyphenyl)-phosphonium compound **1** was used as a hapten to elicit these antibodies (Fig. 1). It was anticipated that the positively charged phosphonium moiety would resemble a positively charged intermediate or intermediates produced from a trityl substrate during an acid-catalysed detritylization reaction. Thus, antibodies specific for the phosphonium hapten were expected not only to bind analogous trityl substrates, but also to stabilize a positively charged intermediate(s) in the cleavage reaction by virtue of charge complementarity in the binding pocket. Since initial protonation of the trityl substrate and subsequent formation of a trityl cation both involve a positive charge, it was not clear whether one or both of these steps would be influenced by the antibody.

The hapten **1** was synthesized by the reaction of tris(4-methoxyphenyl)-phosphine with 7-bromoheptanoic acid. This hapten was coupled to keyhole limpet haemocyanin using a water-soluble carbodiimide reagent ([1-ethyl-3-(dimethylamino)-propyl]carbodiimide) and *N*-hydroxysulphosuccinimide. Interestingly, higher loading of hapten onto the carrier protein produced

TABLE 1 Michaelis–Menten parameters for the cleavage of a variety of substrates by monoclonal antibody 37C4

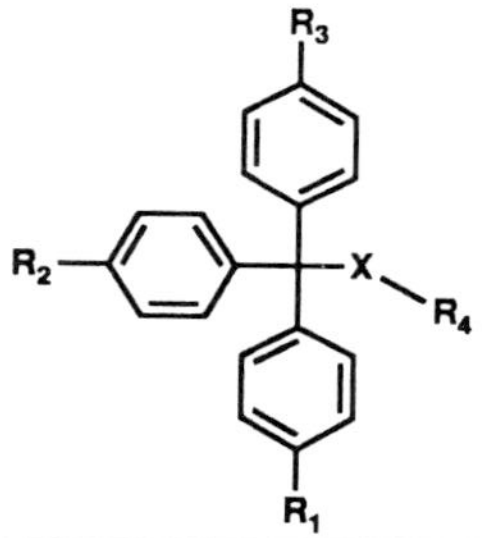

Compound	R_1	R_2	R_3	X	R_4	k_{cat} (min^{-1})	K_m (μM)	k_{cat}/k_{uncat}
2	OCH_3	OCH_3	OCH_3	O	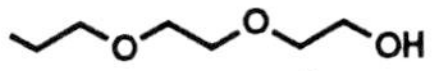	0.1	31	270
3	OCH_3	OCH_3	H	O	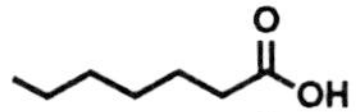	0.072	30	200
4	OCH_3	H	H	O	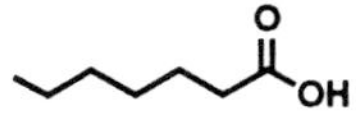	4.2×10^{-3}	30	230
5	OCH_3	OCH_3	OCH_3	S	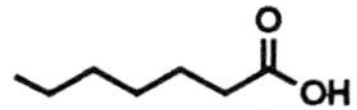	2.2×10^{-3}	33	78
6	H	H	H	N	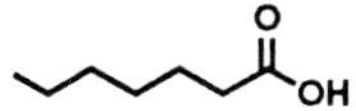	No significant rate acceleration		
7	OCH_3	OCH_3	H	O	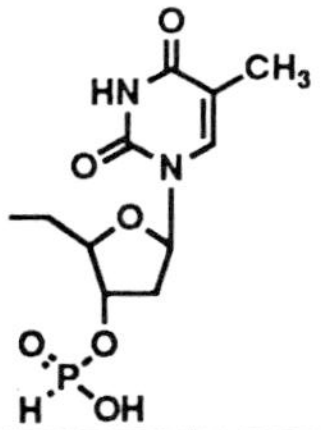	No significant rate acceleration		

Reactions were performed at 25 °C in 25 mM phosphate buffer pH 6.0.

a conjugate which was less antigenic than that made with relatively low loading levels.

Eleven hybridoma cell lines were identified which secreted antibody specific for hapten **1** as judged by ELISA (enzyme-linked immunosorbent assay). Eight of the eleven hybridoma cell lines produced antibodies which displayed significant catalytic behaviour with the 4,4′,4″-trimethoxytrityl ether substrate **2** (see Table 1). The reactions of one of these antibodies, called 37C4, were analysed in detail by high performance liquid chromatography.

Substrate specificity of antibody 37C4

We examined the initial rates of reaction of antibody 37C4 with a number of trityl substrates. In all cases, saturation kinetics was observed; the relevant parameters are listed in Table 1. The catalytic rate enhancement due to the antibody, measured as k_{cat}/k_{uncat} was consistently near 200; the K_m value was close to 30 μM for the 4-monomethoxytrityl, 4,4′-dimethoxytrityl and 4,4′,4″-trimethoxytrityl substrates **2–4**. The thioether compound **5** displayed a lower rate enhancement of 78. The compounds with no methoxy substituents on the trityl group **6** or with steric bulk **7** were not accepted as substrate by 37C4. Taken together, these results indicate that for 37C4 a single methoxy substituent at the 4 position of one of the phenyl rings of a trityl substrate is necessary and sufficient for catalysis to occur. Compound **6** was not even a competitive inhibitor of the reaction of 37C4 with compound **2**, so the lack of reaction observed with **6** probably reflects a lack of binding.

A dissociation constant for hapten **1** of 25 nM was measured with antibody 37C4 using competition ELISA. Consistent with this tight binding, addition of one equivalent of **1** per antibody binding site to the reaction mixture totally inhibited catalytic activity. The product trityl alcohols also inhibited the reactions with 37C4. A dissociation constant of 80 μM was measured for 4,4′,4″-trimethoxytrityl alcohol.

Since the trityl cleavage reaction presumably involves a trityl cation, it is possible that this cation alkylates the antibody binding pocket during the reaction and therefore interferes with the catalysis. To investigate this, we monitored the catalytic reaction through 21 turnovers per binding site (Fig. 2). No decrease in rate was detected, so it appears that this type of antibody alkylation does not occur to a significant extent with 37C4.

Possible mechanism for the cleavage of trityl substrates by antibody 37C4

The catalytic cleavage rate of substrate **2** by antibody 37C4 was determined over the pH range 5.0–8.0 (Fig. 3). The pH-dependent cleavage rates for substrate **2** in the absence of antibody are plotted for comparison. The background and antibody-catalysed reactions display a similar logarithmic increase in rate

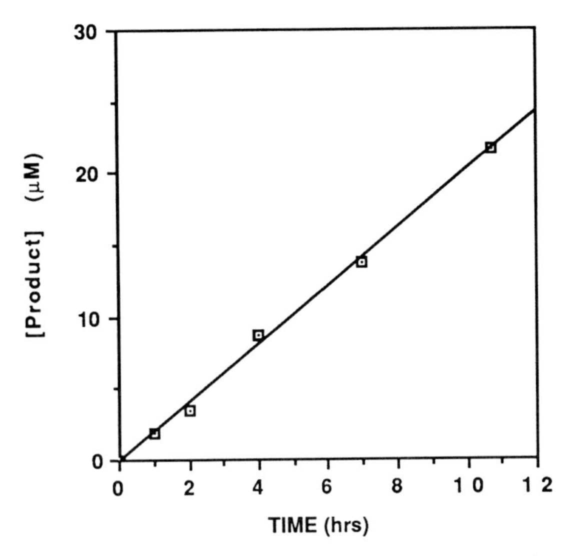

FIG. 2. Plot of the amount of product observed versus time using 100 μM substrate **2** and 0.5 μM antibody 37C4. The reactions were carried out at 25 °C in 25 mM phosphate buffer pH 6.0. The catalytic rate remained constant through all 21 turnovers per binding site.

with decreasing pH. This finding is in contrast to other studies of catalytic antibodies elicited by cationic haptens, which indicated that residues with ionizable side chains did participate directly in the catalytic reaction (Shokat et al 1989, Janda et al 1990). Further studies are being conducted to eliminate the possibility of general acid catalysis in the binding pocket of 37C4.

Geometric distortion of the substrate brought about by binding to the antibody is not likely to be the primary cause of the observed catalytic rate enhancement. The phosphonium hapten and trityl substrates are most probably rigorously tetrahedral around the central atom by analogy with similar structures determined by X-ray crystallographic analysis (Czerwinski 1986, Voliotis et al 1986). The high energy trityl cation, on the other hand, is presumably planar (Gomes de Mesquita et al 1965), so it is not reasonable to suppose that the antibody is effecting catalysis by preferentially stabilizing the geometry of this intermediate.

The positively charged phosphonium hapten would be expected to induce an antibody binding pocket which accommodates this positive charge with some degree of negative charge. Therefore, the best explanation for the catalytic rate increase observed with 37C4 is that charge complementarity in the antibody binding site stabilizes a positively charged intermediate(s) in the cleavage reaction relative to the aqueous medium. This charge complementarity could be the result of specific alignment of dipoles in the binding site or of a negative charge

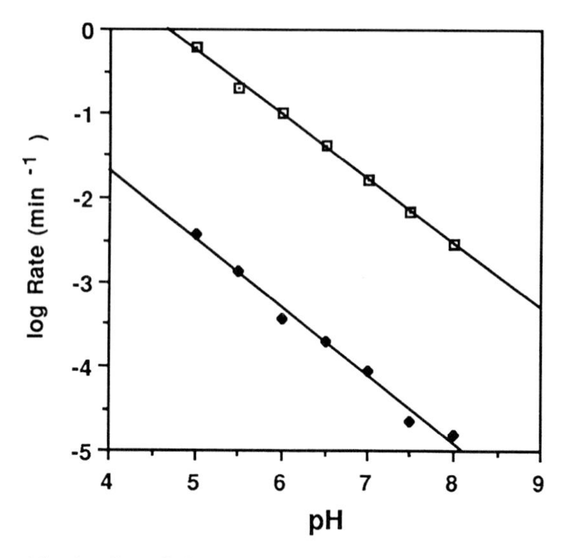

FIG. 3. Logarithmic plot of the pH rate profiles for the reaction of **2** with (as k_{cat}, upper trace) and without (lower trace) monoclonal antibody 37C4. The drawn lines are the best linear fit of all points. Reactions were run at 25 °C in 25 mM phosphate buffer at the given pH. The ionic strength was not held constant throughout the pH range.

provided by an ionized side chain with a pK_a lower than 5. If the above mechanistic interpretation is correct, the 270-fold rate enhancement observed with 37C4 can be viewed as a crude estimate of the catalytic benefit derived from charge complementation in a protein pocket alone; that is without contributions from other factors such as geometric distortion.

Modification of antibody 37C4 by molecular biology

The catalytic rate increase seen with 37C4 is not yet high enough to be useful in a synthetic strategy involving the cleavage of trityl groups. However, an appropriately placed general acid in the antibody binding pocket might increase catalytic activity, since the pH rate profile in Fig. 3 indicates that protonation is the rate-determining step with 37C4. We have therefore produced a novel vector derived from phage λ with the intention of cloning sequences coding for the binding site of 37C4 into microorganisms to facilitate site-directed mutagenesis of the binding pocket and ideally to introduce the desired general acid group. Site-directed mutagenesis has been used to improve the rate observed with another catalytic antibody (Baldwin & Schultz 1989).

Our phage λ vector is different from those previously reported (Huse et al 1989) in that the cloned sequences correspond to the antibody Fv region only, and the light chain is connected to the heavy chain via a linker peptide analogous to the known single-chain antigen-binding molecules. The design and other uses of this new vector will be reported elsewhere.

The mRNA from the 37C4 hybridoma was amplified using the polymerase chain reaction and a variety of primers designed to amplify Fv region sequences only. Double-stranded DNA with the expected length of 300 bp was the primary product in several of these reactions. The amplified sequences are currently being placed in the new vector to be analysed for catalytic activity in preparation for the site-directed mutagenesis studies.

Summary

An antibody has been produced that is capable of catalysing the selective cleavage of certain trityl ether and trityl thioether protecting groups. Catalytic rate enhancements of up to 270-fold were observed. A general acid group in the antibody binding pocket might significantly increase the catalytic efficiency. For this reason, a novel single chain vector based on phage λ has been constructed. The binding site sequences from our tritylase antibody are being incorporated into this vector and site-directed mutagenesis studies will be done.

References

Baldwin E, Schultz PG 1989 Generation of a catalytic antibody by site directed mutagenesis. Science (Wash DC) 245:1104–1107

Czerwinski EW 1986 Structure of antischistosome compounds. I. (2-Aminoethyl)triphenyl-phosphonium bromide hydrobromide. Acta Crystallogr Sect C Cryst Struct Commun 42:236–239

Gomes de Mesquita AH, MacGillavry CH, Eriks K 1965 The structure of triphenylmethyl perchlorate at 85 °C. Acta Crystallogr 18:437–443

Huse WD, Sastry L, Iverson SA et al 1989 Expression of the antibody repertoire in phage λ. Science (Wash DC) 246:1275–1278

Iverson BL, Cameron KE, Jahangiri GK, Pasternak DS 1990 Selective cleavage of trityl protecting groups catalyzed by an antibody. J Am Chem Soc 112:5320–5323

Janda KD, Weinhouse MI, Schloeder DM, Benkovic SJ, Lerner RA 1990 A bait and switch strategy for obtaining catalytic antibodies with acyl transfer capabilities. J Am Chem Soc 112:1274–1275

Lerner RA, Benkovic SJ 1990 Observations in the interface between immunology and chemistry. Chemtracts Org Chem 3:1–36

Schultz PG, Lerner RA, Benkovic SJ 1990 Catalytic antibodies. Chem Eng News 68:26–40

Shokat KM, Leumann CJ, Sugasawara R, Schultz PG 1989 A new strategy for the generation of catalytic antibodies. Nature (Lond) 338:269–271

Voliotis S, Cordopatic P, Main P, Nastopoulos V, Germain G 1986 Structures of dipeptides trityl-glycyl-L-phenylalanine benzyl ester and *tert*-butyloxycarbonyl-L-phenylalanyl-S-ethyl-L-cystein dimethylamide. Acta Crystallogr Sect C Cryst Struct Commun 42:1777–1780

DISCUSSION

Shokat: Have you tried a trityl chloride tritylation reaction? Perhaps you didn't put enough 'spinach' on the linker.

Iverson: The first time I screened these antibodies I used 5% acetonitrile solution and I saw very little hydrolytic activity with any of the antibodies. Using an ELISA I found that with even 20% acetonitrile there was normal binding of the antibody to the hapten. This suggests that high concentrations of organic solvents may not be denaturing the antibody, and the decrease in hydrolytic activity in organic solvent might be due to something else. In the presumed hydrophobic pocket of the antibody the organic solution may be simply replacing the water molecules and preventing them from participating in the acid-catalysed hydrolysis. Therefore, it may be possible to run a tritylation reaction with these antibodies in organic solvent without worrying about competing hydrolysis.

The maximum solubility of the polyether compound **2** in 1% acetonitrile is about 250 μM; the solubilities of the carboxylate compounds **3–5** were a bit higher but I think there was some micellar-type structure. I always worked in the 100 μM range where they were fairly soluble.

A key consideration of these reactions is product inhibition. In this case, where the product looks very like the hapten and substrate, that's a grave concern. But if only one methoxy substituent is required on the trityl group, one could put other things, such as large hydrophobic moieties, on the other two phenyl rings. This should make the product trityl alcohols much less soluble after being cleaved and remove them from solution via precipitation.

Lerner: Can you exploit changes of propeller twists in these antibodies by virtue of what you substitute?

Iverson: I don't know.

Hansen: It would be interesting to look at the ammonium compound and see how subtle changes in bond length affect binding, and what K_i one would observe for that compound.

Iverson: Unfortunately, the monomethoxy amine compounds weren't stable enough to characterize, so I had to use the no-methoxy trityl compound **6** and that didn't bind.

Blackburn: The fact that you can put different substituents on the aromatic groups allows you to explore whether this reaction proceeds with inversion, retention or scrambling of stereochemistry, which will be interesting.

From CPK molecular modelling of whether or not the phosphorus will adopt planarity, it is well established that there's a strong propensity for phosphorus to adopt a valency of five (a) if there is relief of strain; (b) if there is a five-membered ring involved in formation of a transient intermediate, and (c) if the phosphorus thereby gets additional oxygen ligands, especially if it is short of these. I predict that if you put an *o*-hydroxymethyl on to any of those three rings, it would snap shut and you would have a five covalent phosphorane with relative planarity of the three aryl groups. You couldn't guarantee that they would all be equatorial; one might go axial, two would certainly be equatorial. But it would be a 25-step synthesis!

Green: The catalytic activity shown by all eight of your antibodies was about the same. You showed the difference in structure between the hapten and the substrate on the one hand, and between the hapten and the trityl cation on the other, and suggested that considerable flexibility is possible within the binding site. This is an important point: how much motion can take place within the antibody combining site during a chemical reaction? The *cis/trans* isomerization also suggests appreciable motion within the binding site. An alternative interpretation is that the antibody is simply not binding the entire substrate molecule tightly: one part may be rigidly restricted while another part may be relatively free to move. You might be able to check this in your system by studying binding to mono-(*p*-methoxyphenyl) or di-(*p*-methoxyphenyl) derivatives. Do they inhibit binding?

Iverson: I haven't looked at those.

Lerner: Do you think the antibody would recognize the triphenyl phosphine?

Iverson: It may.

Schwabacher: As you removed the methoxy groups, the compounds were all bound by the antibody with about the same affinity. The only one without any methoxy groups was the tritylamine and that is charged.

Iverson: That is not charged, otherwise it would be hydrolysed immediately.

Schwabacher: Presumably, you looked at that rather than the ether because the ether was too slow. Did the ether inhibit?

Iverson: I haven't tried that yet.

Final general discussion

How good are catalytic antibodies?

Jencks: What is the largest rate acceleration that has been reported for catalytic antibodies?

Janda: One millionfold for one that catalyses hydrolysis of an ester that was described by Richard Lerner's group and also for antibody NPN43C9 that hydrolyses the *p*-nitroanilide amide ester that I made and Steve described (Benkovic et al, this volume). However, the antibody that catalyses the bimolecular ester formation reaction that I described earlier (p 37) will easily surpass these.

Paul: With our naturally produced antibody to vasoactive intestinal peptide, there is no background rate of hydrolysis, so the acceleration could be far greater than one millionfold (Paul et al, this volume).

Jencks: To what extent can we identify the factors that are important for that millionfold rate acceleration?

Benkovic: If we use the ratio of the K_i for transition state analogue binding to the K_d for binding of the substrate to predict a ratio for the rate of the spontaneous reaction to that for the antibody-catalysed reaction, we can readily account for the rate acceleration of 10^3 seen in most examples of antibody catalysis.

Jencks: The rate acceleration is accounted for by the geometry?

Benkovic: Yes. I think the extra factor of 100 or even 1000-fold rate acceleration is due to active participation by a side chain in the antibody binding site in catalysis. This is apparently true for the NPN43C9 antibody. It's probably also true of the initial Tramontano/Lerner antibody, because that was stoichiometrically inhibited by an activated ester, suggesting the presence of a highly reactive functional group.

Lerner: The antibody that catalyses the transesterification reaction Kim Janda described should have a very high effective molarity and may be the most efficient catalytic antibody we have seen so far. That's why it is important to get the second order rate constant of the background reaction.

With catalytic antibodies it is not uncommon to get two extremes of catalysts—antibodies that give a rate acceleration of several hundredfold and those that give one million or better. We don't seem to get rate accelerations between those.

Shokat: The *cis/trans* isomerase antibody that Peter Schultz mentioned (p 88) is in the middle: it gives a rate acceleration of about 10^4.

Benkovic: We have to dissect these reactions more carefully. In most cases we are just comparing the rate constants of the spontaneous and antibody-catalysed reactions. We may not be comparing the correct rate constants: chemistry may be rate limiting in one case and product loss in another.

Lerner: What, in general, was the experience in the model systems, like the cyclodextrins and the crown ethers? There was some dispute about the rate accelerations in those systems.

Benkovic: In those debates there was doubt about whether the comparisons were rigorous but we know how to do them now. As expected, the rate accelerations for those systems vary considerably. Some supramolecular complexes give rate accelerations of only 100-fold or less.

Hilvert: The rate accelerations achieved by cyclodextrins can be quite substantial. Some of the acylations, for example, are extremely rapid; rate enhancements can approach 10^8-fold (Breslow et al 1983).

Benkovic: But they don't turn over. I suspect that if we measure the actual acylation rate catalysed by some of the antibodies, particularly those that are stoichiometrically inhibited, they are that fast.

Lerner: I was not asking for comparisons of rates. I was asking whether we are repeating the erroneous history of other systems or have we learned from them.

Benkovic: In the sense that we characterize antibody catalysts merely by k_{cat} and K_m, we could be making the wrong comparisons. We need to look at various classes of reactions to broaden the field before we look at the details of any particular antibody.

Jencks: Is it not important to examine some particular reactions in sufficient detail that one can begin to talk sensibly about the catalytic processes?

Benkovic: That fits with my philosophy, but we should also continue to widen the scope. It is going to be very difficult with condensation reactions that lead to smaller products, in terms of global steric features. To catalyse bimolecular reactions successfully with antibodies such that there is a large number of turnovers is a challenge. We should also carry out detailed mechanistic analysis of antibodies that catalyse reactions efficiently.

Green: The catalytic antibodies described so far have given widely different rate enhancement values. Do certain haptens or certain conditions tend to give higher rate accelerations? How different are the catalytic antibodies from a given batch or a given fusion?

In the example that I described (p 138–141) all four antibodies gave roughly the same rate acceleration. Using the same hapten and screening procedure in a different fusion, more catalytic antibodies were obtained whose k_{cat}/k_{uncat} values were the same as those of the initial four.

Lerner: Are those four antibodies all different?

Green: Yes.

Benkovic: In the lipase series that we worked on with Kim Janda, we had 18 antibodies that were practically all reactive. The range of rate accelerations was from a hundredfold to almost 100 000-fold. We have made the same types of transition state analogues for other cases and produced similar numbers of antibodies, none of which were reactive.

Hilvert: This just points out that a statistically insignificant number of experiments has been done. This is the range of results that one can expect in general.

Burton: If all the antibodies are taken from the same fusion, they may be very similar antibodies.

Hilvert: In our chorismate mutase system, we identified 45 monoclonal antibodies that bind the transition state analogue (Hilvert, this volume). Two of these were catalytic, but kinetic studies and preliminary cDNA sequencing data showed that these molecules are identical. We don't yet know how similar the non-catalytic antibodies are to one another, but the diversity of the antibody population from a single fusion is clearly less, and may be significantly less, than the total number of monoclonals identified.

Burton: We should do multiple fusions or use the λ phage technology, or even look at different animals— individual mice might produce different ranges of antibodies.

Plückthun: The whole concept relies on the hope that one can induce the immune system to supply nucleophiles. This will happen occasionally, but I am very sceptical whether there are general strategies for achieving this without resorting to protein engineering. The question is how to achieve optimal binding. Perhaps one should decrease the binding affinities, e.g. by using substrates not too similar to the immunogen, and not rely on too high binding energies.

Jencks: For the ground state or the transition state?

Plückthun: For the ground state.

Jencks: With enzymes it is essential that ground state stabilization is not too large.

Shokat: We have developed general strategies for eliciting catalytic groups and we see normal Michaelis–Menten kinetics. Clearly, the problem that Andreas Plückthun foresees is not arising.

Plückthun: The basic problem is that enormous rate enhancement will require functional groups and chemical catalysis in addition to what I would call physical catalysis, which is basically a super solvent effect that the protein provides. Are there general strategies to achieve this or do we have to depend on luck that a particular mouse has genes that encode a nucleophile at a suitable position in the framework region of the antibody?

Shokat: What does luck have to do with it? We had a mouse in Maryland that made an antibody that catalysed a β-elimination reaction (Shokat & Schultz, this volume) and a mouse in Berkeley that made an antibody that catalysed a *cis/trans* isomerization. In both cases we elicited specific catalytic groups in the antibody combining site, rationally. Luck had nothing to do with it.

Plückthun: I am sure that the initial screening of the immune system will have to be followed by random mutagenesis.

Lerner: I am not so sure. First, we don't need nucleophiles for all reactions. Furthermore, there is the combinatorial methodology, which is mutagenesis in one sense.

Plückthun: The combinatorial approach really repeats what the immune system already does.

Lerner: With one enormous exception—the ability to introduce into the pocket of the antibody functionalities that are not encoded by the genetic code.

Plückthun: That's exactly what I meant by random mutagenesis.

Hilvert: A single nucleophile or a single general base may not be enough. You may need a constellation of multiple catalytic groups that act synergistically. The probability of such groups arising spontaneously in an antibody is very very small. To get a carboxylate in a binding site that interacts with a histidine which in turn interacts with a serine is going to be extremely difficult.

Lerner: I believe zinc is about as good as a catalytic triad.

Hilvert: But the metalloenzymes, carboxypeptidase and thermolysin, have additional catalytic groups within their active sites; they have a carboxylate that acts as a general base to deprotonate the enzyme-bound water, and they have an arginine to stabilize the anionic transition state or the tetrahedral intermediates. In carboxypeptidase there is a phenylalanine that folds over the active site: it doesn't play a role as a general acid but it seems to be required, perhaps as a hydrophobic lid to cover the substrate and exclude water from the active site.

Shokat: When you remove one of two carboxylates from a β-glucosidase it still gives a rate acceleration of 10^6. That is only three orders of magnitude lower than the wild-type. Most of us would be happy with a glucosidase catalytic antibody that gave a rate acceleration of 10^6.

Hilvert: It's great that we can get rate accelerations of 10^6 using antibodies; but can we get 10^{14}?

Plückthun: We should remember the very important experiments that Jim Wells did on subtilisin, where he castrated the enzyme by taking out the catalytic triad and the rate enhancement of about 1000-fold resembled what is seen for most catalytic antibodies (Carter & Wells 1988, 1990). This will be the general case, and this is what is usually observed. Some people may be lucky and find a catalytic group by chance but I am very pessimistic about this as a general strategy.

Jencks: How should it best be done?

Plückthun: These antibodies should be taken as starting points: experience will show whether random mutagenesis or structure determination followed by protein engineering will get further.

Blackburn: Are statistics for random mutagenesis more favourable than picking one of 10^8 antibodies?

Plückthun: Yes, if you can focus the mutagenesis on certain regions and if you have a good selection system. The attractive point is that you can do the mutagenesis in multiple stages: if the probability of finding something is one in a million in three sequential reactions, it is feasible, whereas finding something in 10^{18} is difficult.

Benkovic: If by mutagenesis you construct an antibody that achieves the same rate acceleration as an enzyme, it will be fascinating to see if the antibody has the same active site, in terms of the positioning of functional groups, as the enzyme. We might find out whether there are a number of possible pathways of comparable free energy for a given reaction.

The odds against switching from one pathway to another by protein engineering are high. We haven't been able to convert subtilisin to papain, even though the geometries of the active sites are comparable. Substitution of cysteine for sulphur blocked the active site, which was not expected. This refers to the point of catalysing reactions that enzymes don't. We should find catalysts for reactions that chemists can't do very well either.

Burton: Natural selection for antibodies is always on binding, not catalysis.

Paul: Is that true? There is a finite binding affinity that antibodies achieve in Nature—a K_d of 10^{13} or 10^{14}: most mature antibodies have an even lower affinity. An upper limit on the affinity could be imposed by the rate of somatic mutation of the complementarity-determining region. Alternatively, something may turn off the lymphocyte if the antibody it is making binds with too high affinity. B lymphocytes expressing high affinity antibodies are tolerized by antigen, probably by receptor desensitization and down-regulation (Goodnow et al 1990). If a catalytic antibody could be formed by somatic mutation, it would confer selective advantage on the cell because the surface-bound antigen could be removed.

Burton: Then why are the effector mechanisms there? Why is there an Fc if the immune system relies just on cleaving?

Paul: There are many teleological arguments in favour of catalysis by the antibody active site. The Fc fragment is needed to interact with the Fc receptor on cells and complement components. Neither of these effector mechanisms is antigen specific and each requires the participation of several molecules and organized structures. Catalysis by antibody active sites would be an independent means of providing rapid, specific, efficient immunological defence. The real question is: How general is this phenomenon? Going back to antibody affinity, it should be recognized that there is a point beyond which affinity does not increase.

Burton: The body has no need for very high affinities, probably not greater than 10^9.

Hansen: To go back to the question of binding the transition state analogue—are antibodies that bind transition state analogues most strongly the best catalysts? Since the transition state analogue is only an analogue, should one search for antibodies that bind it weakly and are more complementary to the true transition state?

Janda: Typically, we select for antibodies that have the highest titre. You also have to keep the hybridoma cells alive. We try to get a range of different antibodies, but if you start with 200, you may end up with 40 because of attrition. So although we screen for the antibodies that bind most tightly to the hapten, we may not end up with them. We don't go after the weakest binders but I think we get a whole array of different binders.

Hansen: Within that array, does the antibody with low binding affinity have higher catalytic activity?

Janda: By ELISA you can't quantitate that.

Jencks: Affinity in its simplest sense is not what you would like to rely on to get catalysis; you want other groups in the antibody binding site that will contribute chemical catalysis. In some antibodies there are groups that bring about chemistry, such as proton transfer, but I am not sure how one screens for those.

Harris: This goes back to Andreas Plückthun's point. Were you thinking of bringing in modified light chains by cassette mutagenesis of the complementarity-determining regions?

Plückthun: That's one way; there are many ways of doing this and it's a question of which techniques will prove to be easier than others.

Green: It also depends on the application. In many cases it is not rate enhancement that is critical but specificity. This is especially true for stable chemical bonds. For example, subtilisin to enzymologists looks 'dead' when its catalytic triad has been replaced by three alanines and its activity has decreased by a factor of about 10^6. But for a chemist the residual activity of this mutant (a rate enhancement of about 10^3) is still extraordinary for a peptide bond hydrolysis (Wells & Estell 1988).

Lerner: A 100-fold rate enhancement converts to about 98% enantiomeric excess.

Page: What is the highest number that's been obtained for enantiomeric excess or discrimination between enantiomers using catalytic antibodies?

Benkovic: For the antibody that catalysed the lactone cyclization, we used NMR and a europium shift reagent and measured an enantiomeric excess of about 98%. We plan to do a more stringent analysis, but I have no doubt that the enantiomeric excess will be greater than 99%. The glory of antibody systems is that such specificities are inherent.

Green: In 1928 Landsteiner & van der Scheer showed that a polyclonal antibody preparation raised against one enantiomer of a phenylglycine derivative had a stereospecific preference for that same enantiomer.

Page: That was binding discrimination. We are talking about kinetic resolution in terms of discrimination between two substrates.

Hilvert: We have done a kinetic resolution using the chorismate mutase antibody. Of course, one doesn't need catalysis to discriminate enantiomers;

one could just run the racemate down a column of a non-catalytic antibody that recognizes only one optical isomer.

Paul: Could there be different residues in an antibody active site responsible for tight binding and for catalysis? That is probably the only way to exploit the ability of an antibody to discriminate between related substrates and still get a good catalyst. If the binding energy of a catalytic antibody is concentrated at the site of the chemical reaction, it will lose specificity. If the binding involves too many remote sites, less binding energy will be available for transition state stabilization.

Jencks: To what extent is binding specificity related to catalytic efficiency in these antibodies?

Lerner: Everything we have heard suggests there is a fairly strict relationship, for example the handedness of the lipases (Janda p 37) and the number of methoxy groups on the trityl groups (Iverson et al, this volume).

Janda: I disagree. If we make a rough estimate of binding specificity using the Michaelis constant (K_m) and compare this to catalytic efficiency, in our antibodies that mimic lipases, nine were enantiospecific for R isomers, two for S isomers. All those antibodies had different K_ms: one gave a very large rate acceleration and had a high K_m; another had a moderate K_m and the rate acceleration was rather poor.

Hilvert: I don't think that you can conclude anything at all about binding from a K_m value, unless you know the details of the mechanism.

Page: Some of the examples we have heard about at this symposium have had selectivity expressed in terms of k_{cat}.

Benkovic: Antibody NPN43C9 has a measured K_d for the *p*-nitrophenyl ester of about 1 μM. It is inhibited by the *m*-nitro compound, which is not a substrate, with about the same K_i: it is a competitive inhibitor so that's also a K_d. This is an example of no difference in selectivity but a difference in reactivity.

Jencks: So these antibodies behave in the same way as enzymes: specificity may be in binding or in catalysis or both, but there's no necessary correlation between them.

Paul: They may behave like enzymes because they are raised against transition state analogues. There might be some phenomena that are unique to antibodies which would be seen only when you worked in the non-mimic situation.

Suckling: What chemistry do antibodies have that enzyme proteins do not have?

Paul: The crucial difference is the size of the active site. The largest protease active site in the literature is that of renin, which can accommodate eight amino acids of the substrate (Sibanda et al 1985). In antibodies, the recent data say that 15–22 amino acids of the antigen can make contact with the antibody active site (Davies et al 1990). Some of those are energetically favourable, some are less favourable.

Suckling: But that is just having more of the same, not introducing new chemistry.

Paul: The specificity manifestations could be different: there could be remote effects in an antibody molecule that you could not have in an enzyme because of the size of the active site.

Wilson: But you are not dealing with the whole binding site in hapten–Fab interactions; you may be dealing with a very small cavity in which the hapten is almost completely buried, so it can be like an enzyme active site.

Paul: In the case of haptens, perhaps yes, because the hapten is a small molecule. When you go to large polypeptides where there are going to be large areas of contact, then the situation is likely to be different.

Wilson: We have crystal structures of two antibodies bound to peptides. In one, only seven of 19 amino acids in the peptide contact the antibody (Stanfield et al 1990).

Plückthun: There may be more contacts that are disordered so you can't see them.

Wilson: But they presumably do not contribute any binding energy.

Potential uses of catalytic antibodies

Jencks: What do people think will be the usefulness of catalytic antibodies in science and industry?

Suckling: They are a promising alternative technology to known methods for generating new protein-based catalysts—microbiology, modified enzymes and so on. An important question is whether one can generate proteins with well-defined specificities to catalyse key steps in synthetic pathways so that optically pure homochiral products can be obtained.

Jencks: Is it feasible to get large enough quantities of catalytic antibodies to do that kind of thing?

Suckling: That's something we have to prove but I feel it ought to be.

Harris: You have to look carefully at the time scales when deciding whether antibodies as catalysts are a cost-effective alternative to enzymes. For example, consider that a compound of interest can be obtained only by a rather difficult chemical synthesis. The chemists must have made it for you to have found out that it has some useful effect. You have to weigh the time it takes to design a transition state analogue and to elicit an antibody to do a step in the synthesis against the time required to screen microorganisms to look for an enzyme which does that transformation. Meanwhile, the chemists will be finding alternative synthetic routes for that compound.

Lerner: If the drug will be produced commercially for 20 years, the time to make the transition state analogue to raise a catalytic antibody, or the time to screen microorganisms, is worth it.

Suckling: There are a lot of cases where the process makes the money for the company rather than the product. If you are in industry, you look at all the possibilities and take the best. Catalytic antibodies have not yet been demonstrated to be commercially competitive.

Hilvert: Although there was great excitement for using enzymes in industry some years back, over the last 10–15 years relatively few processes have been developed that employ enzymes.

Harris: And there are very few drugs that are made with a biotransformation as part of the process, for all sorts of reasons related to cost effectiveness.

Janda: Is that the fault of the enzyme or the fault of the people in industry who don't want to change?

Harris: I am told that one can usually find ways around difficult synthetic steps.

Paul: You need small, stable catalysts: that's one reason why industry doesn't want enzymes and may not want antibodies.

Hilvert: In addition, if there is existing technology to make a particular molecule, it may be impractical to re-tool the factory to use an enzyme.

Lerner: The bottom line is not the industrial application, it is that you have a programmable binding site. How much can one learn about the potential of a proteinaceous pocket to facilitate chemical transformations? How many different ways are there for proteins to catalyse similar chemical transformations? This touches on the evolutionary issue of the difference between something that didn't happen and something that can't happen. Catalytic antibodies give us the ability to approach that which physical organic chemists have laid on the table for 20–30 years using programmable binding. If catalytic antibodies never got past nitrophenyl esters, they could still teach us a lot, and we are well past nitrophenyl esters. I object to people who say that if you don't make something that industry can use, what good is it. As Churchill said, 'Madam, what good is a new-born baby?' And I would say that the catalytic antibody field is now more of a teenager—which is even less use!

Suckling: Teenagers are also usually non-programmable!

References

Benkovic SJ, Adams J, Janda KD, Lerner RA 1991 A catalytic antibody uses a multistep sequence. In: Catalytic antibodies. Wiley, Chichester (Ciba Found Symp 159) p 4–12

Breslow R, Trainor G, Ueno A 1983 Optimization of metallocene substrates for β-cyclodextrin reactions. J Am Chem Soc 105:2739–2744

Carter P, Wells JA 1988 Dissecting the catalytic triad of a serine protease. Nature (Lond) 332:564–568

Carter P, Wells JA 1990 Functional interaction among catalytic residues in subtilisin BPN′. Proteins Struct Funct Genet 7:335–342

Davies DR, Padlan EA, Sheriff S 1990 Antibody–antigen complexes. Annu Rev Biochem 59:439–474

Goodnow CC, Adelstein S, Basten A 1990 The need for central and peripheral tolerance in the B cell repertoire. Science (Wash DC) 248:1373–1379

Hilvert D 1991 Antibody catalysis of carbon–carbon bond formation. In: Catalytic antibodies. Wiley, Chichester (Ciba Found Symp 159) p 174–187

Iverson BL, Iverson SA, Cameron KE, Jahangiri GK, Pasternak DS, Lerner RA 1991 Tritylase antibodies. Wiley, Chichester (Ciba Found Symp 159) p 227–235

Landsteiner K, van der Scheer J 1928 Serological differentiation of steric isomers. J Exp Med 48:315–320

Paul S, Johnson DR, Massey R 1991 Binding and multiple hydrolytic sites in epitopes recognized by catalytic anti-peptide antibodies. In: Catalytic antibodies. Wiley, Chichester (Ciba Found Symp 159) p 156–173

Shokat KM, Schultz PG 1991 The generation of antibody combining sites containing catalytic residues. In: Catalytic antibodies. Wiley, Chichester (Ciba Found Symp 159) p 118–134

Sibanda BL, Hemmings AM, Blundell TL 1985 Computer graphics modelling and the substrate specificities of human and mouse renins. In: Kostka V (ed) Aspartic proteinases and their inhibitors. Walter de Gruyter, New York, p 339–354

Stanfield RL, Fieser TM, Lerner RA, Wilson IA 1990 Crystal structures of an antibody to a peptide and its complex with peptide antigen at 2.8 Å. Science (Wash DC) 248:712–719

Wells JA, Estell DA 1988 Subtilisin—an enzyme designed to be engineered. Trends Biochem Sci 13:291–297

Summary

W. P. Jencks

Graduate Department of Biochemistry, Brandeis University, Waltham, MA 02254, USA

It is extraordinary that in only four years monoclonal antibodies with high catalytic activity have been discovered, characterized, and studied in several different ways, as we have heard in this symposium. Progress over such a short time is unusual, and the field is still advancing rapidly. It has begun to come of age and we can now ask what will be the most interesting and important developments concerning catalytic antibodies; we have discussed many of them at this meeting.

We have learned a great deal about the structure of catalytic antibody systems, again in an extraordinarily short time. X-ray structures are known for some antibody–antigen complexes and we can monitor conformational changes that occur when small molecules bind to antibodies. Antibody molecules do not just have a site into which an antigen can fit; they can fold over the binding site to increase the available binding energy and perhaps bring about additional stabilization of the transition state. This emphasizes the importance of binding energy for catalysis and the fact that the greater the interaction between enzyme and substrate, the more binding energy is available to destabilize the ground state, to compensate for loss of entropy, and to do whatever is needed to reach the transition state more easily.

Many different classes of reactions have already been shown to be catalysed by antibodies. These cover a considerable fraction of the known classes of reactions, including an elimination reaction in which a carboxylate group that acts as a base has been induced by a cationic ammonium ion; there is a β-elimination reaction that is increased in rate by 10^3–10^4-fold. We have learned about groups that can bring about particular types of catalysis: arginines and, to a lesser extent, protonated lysines act as electrophilic catalysts and as binding sites for anions; carboxylate groups serve as bases and as groups that can bind cationic substituents on a substrate.

Catalytic antibodies are likely to be most interesting when they tell us more about mechanisms of catalysis. We have heard about an entropically driven reaction in the Claisen condensation, which has an entropy requirement of some -13 entropy units; this corresponds rather well with the rate acceleration that is observed with the catalytic antibody. General acid-base catalysis is possible if the appropriate acids and bases can be induced in the active site using immunizing agents that contain charged groups. Finally, changing substrate structure through distortion or strain is probably the oldest proposal for the mechanism of action of enzymes. There are a few cases in which there is evidence for substrate destabilization by strain or distortion in enzyme catalysis, and there are already some indications of it in catalysis by antibodies. We are beginning to get some values for the amount of rate acceleration that can be expected in relatively simple cases. With antibodies we are building up from very simple to more complex catalysis, whereas with enzymes we take an enzyme and reduce it by removing catalytic groups to find out what the different parts of the active site do to accelerate the reaction.

One of the most interesting findings is that it is possible to put metal ions specifically into antibody molecules and to use these to change the chemistry of the reactions that are occurring, in ways that are not easy to do with ordinary amino acid side chains. This provides a powerful direction for developing strong catalysis. We have also heard about metalloenzymes that work by delivering hydroxide ion to the substrate. They lower the pK of water and may interact with other groups to provide electrophilic catalysis.

Clearly, one of the most attractive potentials of catalytic antibodies is to increase the rates of bimolecular reactions. The antibody can bind two molecules, placing them in the correct position to react. The loss of entropy that occurs is compensated by the binding energy and rate accelerations of, in principle, up to 10^8 for 1 M solutions, can be achieved, although a more likely number is between 10^4 and 10^5. There are at least three enzymes for which rate accelerations or stronger binding by factors of the order of 10^4 to 10^5 can be ascribed to induced intramolecularity. So far, effective molarities of up to more than 120 M have been described for catalytic antibodies. In addition reactions, the problem then is to dissociate the single molecule product, which is likely to bind very tightly indeed causing product inhibition. This is less of a problem if the product can be made to 'explode' in some way, or if there is a single transition state that gives two products, as in a bimolecular substitution reaction. This should be a fertile area for further development.

We have heard much about methods for the development and identification of catalytic antibodies, which are clearly both critical problems in the field. Combinatorial antibody expression libraries have been made and will certainly be useful. There are indications that one can use λ bacteriophage vectors to express antibody genes in *Escherichia coli*. These methods are moving ahead very rapidly and will be beneficial for both academia and industry.

We can ask what is likely to happen in the future with catalytic antibodies. I rather hope that I can't predict that! Some likely events are the study and development of antibodies that will catalyse bimolecular reactions having the very large potential for rate acceleration. General acid-base catalysis, and possibly concerted acid-base catalysis, is attractive. Metal ions are particularly attractive, now we know they can be inserted into catalytic sites on antibodies.

It may be possible to bind coenzymes to catalytic antibodies such that they can catalyse reactions that enzymes catalyse using these same coenzymes.

The most obvious goals for further investigation of catalytic antibodies are to obtain quantitative information that will help us to understand the contributing factors in catalysis by enzymes, and to bring about specific chemical changes in large, complex molecules. It is difficult to deal with such labile molecules using biochemical methods without destroying them, and a catalytic antibody that can bring about specific transformations in a particular part of such molecules should have great potential, both in the laboratory and, possibly, in industry, as an effective means of accelerating reactions.

I think we should, therefore, give our very special thanks to Derek Chadwick and to all at the Ciba Foundation who have done an extraordinary job of setting up this very timely, informative and enjoyable symposium.

Index of contributors

Non-participating co-authors are indicated by asterisks. Entries in bold type indicate papers; other entries refer to discussion contributions

Indexes compiled by Liza Weinkove

Subject index